A BRIEF HISTORY OF ALBERT EINSTEIN

Space, Time, and Quantum Theory: $E=mc^2$ and the Foundations of Modern Physics

---

SCOTT MATTHEWS

Contents

"I have no special talent. I am only passionately curious."

- Albert Einstein

Introduction

In the record of history, few names sparkle with the brilliance of Albert Einstein, whose theories redefined the laws of nature and whose life mirrored the tumultuous waves of the twentieth century. *A Brief History of Albert Einstein* delves into the extraordinary journey of Einstein, not just as a physicist who unveiled the mysteries of the universe but as a profound thinker and humanitarian whose impact went beyond the boundaries of science. His story is one of intellectual triumph and personal struggle, set against the backdrop of global upheavals – World Wars, the rise of fascism, and the dawn of the nuclear age.

Einstein's life was as complex as the theories he proposed. Born in the relatively quiet world of a late 19th-century German Jewish family, he evolved into a prominent figure whose thoughts and opinions were sought on a range of issues: from the ethics of atomic power and the philosophy of science, to human rights and the necessity of global governance. His journey from a young, curious mind in Munich to a global icon in Princeton encompasses not only his scientific odyssey but also his relentless quest for peace and understanding in a world that was often on the brink of self-destruction.

As you turn these pages, you will travel through Einstein's life – from his early years marked by intense curiosity and profound solitude to his later years as a global advocate for peace. Through detailed research and insightful analysis, this book aims to showcase how Einstein's groundbreaking theories emerged alongside his deep moral considerations and how his global advocacy for peace and dialogue was formed by his scientific understanding of the universe.

This book is not just a biography but a mirror reflecting the dual nature of human progress – our capacity for great knowledge and great evil. It invites you to reflect on the powerful role of science in shaping human history and the responsibilities of genius in the face of moral and ethical dilemmas. Prepare to be captivated by the story of a man who dared to look into the very essence of nature while never losing sight of humanity's potential for both destruction and greatness. Join me on a journey through the life of Albert Einstein, where the realms of imagination meet the harsh truths of reality, crafting a story as compelling as the mysteries he spent his life unraveling.

Early Years: Family Roots and Childhood

Albert Einstein, a name synonymous with genius and creativity, was born on March 14, 1879, in the pleasantly old-fashioned city of Ulm, in the Kingdom of Württemberg of the German Empire. The Kingdom of Württemberg was a state in the southwest of Germany, part of the German Empire from 1871 until 1918. His parents, Hermann Einstein and Pauline Koch, were not academically inclined themselves but belonged to families that appreciated the arts and sciences. This familial dynamic created an enriching environment for young Albert.

Hermann Einstein, Albert's father, was an enterprising salesman who later co-founded an electrochemical company. Despite his business endeavors not always being successful, Hermann displayed a spirit of resilience and innovation. Pauline, Albert's mother, was an accomplished pianist. Her musical talents were not just a personal joy but also a formative aspect of Albert's upbringing, deeply influencing his love for music throughout his life.

The family's Jewish roots were culturally significant but they were not extremely religious; thus, Albert's early exposure to religious practices was limited. The family celebrated Jewish traditions in a secular fashion, which laid the foundation for Einstein's later views on spirituality and religion.

As a young child, Albert did not exhibit signs of the prodigy he was to become. Chubby, pale, and with thick black hair, he was very reserved and introspective, causing his parents considerable concern about his silence. They even consulted several doctors who reassured them that nothing was wrong with the little boy. It is said that Albert didn't say his first words until he was between three and four years old.

While other boys his age engaged in loud and bouncy play and games like soldiers, Albert was distinctly different. He found the sight of marching soldiers unsettling and preferred being alone, often engaging in thoughtful play with blocks and card houses. The block houses and card houses sometimes reached up to fourteen stories high. Despite repeated visits to doctors due to concerns about his quiet and solitary attitude, no issues were ever diagnosed, and soon it became clear to his parents that Albert was simply a contemplative child, more inclined toward thinking than speaking.

In 1880, the Einstein family relocated to Munich, marking an important chapter in their lives. Upon their arrival, Hermann Einstein and his brother Jakob established Elektrotechnische Fabrik J. Einstein & Cie, a company specializing in the production of electrical devices powered by direct current. This business not only provided the family with momentary financial stability but also introduced young Albert to the wonders of science.

A year after settling in Munich, the family welcomed Albert's younger sister, Maja. Initially, Albert, nearly three years old, viewed his new sibling as a simple uninteresting baby. However, as Maja grew, their relationship grew too, transforming from mere familial bonds into a deep friendship. The siblings became inseparable, sharing many playful moments and adventurous hikes, and exploring the natural surroundings of their home city.

This period of joy and professional success for the family was also when Albert's curiosity about the natural world began to take place. Intrigued by his father and uncle's business dealings with batteries, generators, and wires, Albert developed a fascination with electricity. To him, electricity was an invisible, powerful, and mysterious force that sparked endless questions about its nature, speed, and the potential existence of other unseen forces in the universe.

The initial success of Hermann and Jakob's business endeavors afforded the family a comfortable lifestyle, which enabled them to prioritize Albert's education. At the age of five, Albert was enrolled in a Catholic elementary school in Munich. Despite its religious affiliation, the school was chosen for its strong academic reputation. Here, Albert first experienced formal education, setting the foundation for his future academic pursuits and nurturing his rapidly growing curiosity about the world. This environment, coupled with the stimulating intellectual challenges posed by his family's business, shaped Albert into a thoughtful and inquisitive young student, eager to explore the mysteries of the universe.

Even though Albert was an intelligent child, he was often seen as withdrawn by his teachers. His speech development was delayed, which initially also concerned his teachers. Einstein himself later described this as a phase of thoughtful observation. This characteristic thoroughness became a hallmark of his scientific method. Albert himself noted in various writings that his slow verbal development allowed him to observe and analyze the world differently from others, giving rise to novel insights.

One of the most significant early influences on Einstein was a compass given to him by his father. Regardless of how he manipulated the compass, the needle always pointed north. Albert was intrigued by the invisible force exerted by the Earth, which he learned acted like a giant magnet influencing the compass needle. Young Albert was fascinated by the invisible forces that moved the needle – a curiosity that would eventually lead him to unravel some of the universe's greatest mysteries.

Despite these early signs of brilliance, Albert's academic journey was not without challenges, especially after he had transferred to Luitpold Gymnasium* school at the age of eight. His indifference toward rote learning and strict educational methods often put him at odds with his teachers. The strict regimen of the German school deeply unsettled Albert. The uniformed students marched between classes like soldiers, a sight that made Albert uneasy given his aversion to militaristic order.

In the classroom, strict discipline prevailed; students were expected to sit up straight, respond promptly to the teachers' commands, and refrain from asking questions. This rigid environment stifled Albert's intellectual spirit, as he was expected to memorize information rather than engage with it critically. He disdainfully referred to his teachers as 'sergeants' due to their authoritarian rule over students. This conflict with the educational norms of the time was a catalyst for his later educational philosophies.

---

* Gymnasium (and variations of the word; pl. gymnasia) is a term in various European languages for a secondary school that prepares students for higher education at a university.

However, mathematics served as a sanctuary for Albert, where reasoning was essential and learning extended beyond simple rote memorization. His uncle Jakob, who was an engineer, presented him with complex algebra problems that Albert found thrilling, much like solving intricate puzzles. Albert also taught himself geometry, intrigued by how different shapes fit together, similar to how he played with blocks as a child. This passion for math was further ignited at the age of twelve when he discovered a geometry book he affectionately termed his 'holy little geometry book.' This book was a turning point in Einstein's intellectual journey, as it deepened his engagement with math. Even as he faced punishment at school for his persistent curiosity, the knowledge in the book contrasted sharply with basic arithmetic that was taught in school.

At the age of thirteen, Einstein's interests expanded to encompass music and philosophy. Despite initial resistance as he did not enjoy violin lessons as a child, his perspective transformed dramatically at the age of thirteen upon discovering Mozart's violin symphonies. This experience sparked a lifelong love for Mozart's compositions and a more enthusiastic engagement with music.

At thirteen, Albert was also introduced to the works of Immanuel Kant, specifically the "Critique of Pure Reason." Afterward, Kant quickly became his favorite philosopher. Einstein's tutor noted that although he was only thirteen, he could understand the complex ideas of the philosopher Kant, which are usually incomprehensible to most people.

Albert never quite made good friends with his peers. He had little interest in sports and found the classroom exceedingly dull. What he needed was a mentor, someone who could reassure him that he would overcome these challenging years. That mentor came in the form of Max Talmey, a medical student and a frequent guest at Einstein's dinner table. Max recognized Albert's brilliance early on and supplied him with books from the university, broadening his intellectual horizons. Although Max could not keep pace with Albert's advanced mathematical thinking, he encouraged Einstein's exploration in other fields such as history and religion.

Albert's family, while Jewish, did not strictly observe all Jewish customs, but like everything, Einstein also became interested in understanding religion. At one point, Albert chose to practice Jewish dietary laws more rigorously by abstaining from pork, although he still questioned the literal interpretations of the Bible. His religious curiosity was never about forceful belief and practices but about exploring broader existential questions. Albert believed that "ideas come from God," and he often expressed that his scientific pursuits were driven by a desire to "read God's mind."

In 1894, Hermann and Jakob Einstein's company faced a critical setback when they failed to secure a contract to install electric lighting in Munich. The project required a transition from direct current to the more efficient alternating current, but the company lacked the necessary capital for such technological updates. This failure prompted the sale of their Munich factory and led the family to seek new opportunities in Italy. They initially moved to Milan and shortly thereafter settled in Pavia, at the Palazzo Cornazzani.

During this tumultuous period, Albert Einstein, then fifteen years old, stayed behind in Munich to complete his schooling. Although his father hoped he would pursue electrical engineering, Albert found the rigid structure and teaching methods of the Gymnasium very stifling. The educational environment, which discouraged questioning and independent thinking, deeply frustrated him. This frustration was doubled by the departure of his mentor, Max Talmey, to America. This loss deeply affected Albert as he had not only lost a mentor but also a friend. His isolation was intensified as his family relocated, leaving him to navigate his final school years alone.

Albert's growing disillusionment with the educational system led to a critical attitude against unquestioning obedience to authority, which he famously criticized as 'the greatest enemy of the truth.' This rebellious attitude resulted in conflicts with his teachers, who labeled him a disruptive influence and a "lazy dog" for his challenging questions that often went unanswered. Einstein became increasingly disillusioned with the rigid educational system at the Gymnasium. Frustrated by its strict regimen and lack of creative freedom, he decided to leave the school before completing his education. Seeking a fresh start and a more supportive environment, he moved to Italy to join his family, who had relocated there earlier.

Albert soon joined his parents and sister in northern Italy, a country markedly different from Germany. He was immediately taken by the warmth and open-mindedness of the Italian people. Over the next two years, he immersed himself in the culture, attending concerts and exploring art museums. This period also allowed him ample time for reading and contemplation. He delved into the biographies of scientists Nicholas Copernicus and Galileo Galilei who, like him, had faced criticism and persecution for their revolutionary ideas. These readings further catalyzed his own thinking.

In Italy, Albert dedicated himself to thoroughly documenting his thoughts and exploring the scientific questions that had long captivated him. This period was crucial as it marked his serious commitment to becoming a scientist. During this reflective time, he deepened his understanding of electricity and magnetism, which inspired him to draft his first scientific paper on the subject. Although this paper, titled "On the Investigation of the State of the Ether in a Magnetic Field," was not published, it represented a significant intellectual endeavor for the young Einstein. In this paper, he boldly challenged the widely accepted concept of the ether – an invisible medium thought to permeate space. This challenge showcased his willingness to question established scientific doctrines and laid the groundwork for his future revolutionary theories. Despite the lack of immediate recognition, this early effort was a vital step in Albert's journey as a physicist, helping to clarify his thoughts and refine his theoretical perspectives.

In Italy, Albert also spent many days hiking in the mountains alone, reflecting on his life and future. These walks were a response to his family's failing business and his feeling of being a financial burden.

The solitude and the rhythm of his hikes sparked his creative mind. It was during these reflective moments that Albert decided to pursue physics at the university level and aim for a career as a physics professor. He resolved that no educational institution would ever dominate his intellect as his German high school had attempted to. He also concluded that the freedom to think and explore his own ideas would always be the most important thing in his life. Even if he were to marry and have a family, he understood that his need for intellectual freedom would always take the upper hand. While some people might have viewed this prioritization as selfish, for Albert, it was essential to his very being and future contributions to science.

Recognizing the need for a more formal education to achieve his goals, Albert began preparing to take the entrance exams for the Swiss Federal Polytechnic in Zurich, Switzerland. His preparation was intense, as he was determined to succeed on his own terms. In 1895, at sixteen, slightly younger than usual, Albert sat for the exams. Although he excelled in the mathematics and physics sections, his performance in other subjects was not strong enough to gain admission.

Guided by the recommendation of the principal at Zurich Polytechnic, Albert Einstein decided to complete his secondary education at the Argovian cantonal school, a prestigious Gymnasium located in Aarau, Switzerland. This decision marked an important step in his educational journey, graduating in 1896. During his time in Aarau, Einstein lived with the family of Jost Winteler, a professor who had a profound influence on his intellectual and personal development.

While residing with the Winteler family, Einstein formed a strong bond with them, experiencing a nurturing and intellectually stimulating environment. It was here that he developed a close relationship with Marie Winteler, Jost's daughter, who became his first love. This relationship, although it did not last, played a significant role in Einstein's emotional life during a formative period.

Along with science, Albert Einstein's profound engagement with music was evident during the school in Aarau, Switzerland, especially in a school event where he played Beethoven's violin symphony. An examiner, struck by Einstein's deep emotional connection to the music, described his performance as "remarkable and revealing of great insight." This emotional depth in his musical expression was a rare and highly valued trait, showcasing an aspect of Einstein's personality that resonated deeply with his broader intellectual and artistic sensibilities.

In a momentous move during his stay in Aarau, and with the support of his father, Einstein made the decision to renounce his citizenship of the German Kingdom of Württemberg in January 1896. This action was motivated primarily by his desire to avoid mandatory military service, reflecting his lifelong pacifism* and discomfort with militarism. This renunciation underlined his growing independence and marked a significant step toward defining his personal values and future.

Einstein's academic performance at the Argovian cantonal school culminated in his graduation in September of that year. The graduation acknowledged his academic prowess, as he excelled across most of the curriculum. He received the highest possible grade of six in several subjects including history, physics, algebra, geometry, and descriptive geometry. These grades reflected not only his intellectual abilities but also his capacity to master diverse academic challenges.

Interestingly, Einstein's connection to the Winteler family extended beyond his own educational and personal experiences. His sister, Maja, would later continue the family connection by marrying Paul Winteler, Jost's son. This intertwining of the Einstein and Winteler families highlighted the deep and lasting relationships formed during Albert's year in Aarau, which influenced both his personal life and his emerging scientific perspective.

---

* Pacifism is the belief that violence, including war, is unjustifiable under any circumstances, and that conflicts should be settled through peaceful means. Pacifists often emphasize the use of nonviolent resistance and negotiation to solve disputes, opposing the use of physical force as a means of achieving goals. This ideology can be adopted for moral, religious, philosophical, or practical reasons, with adherents advocating for peace and non-violence in all aspects of human interaction.

These experiences in Aarau were crucial in shaping Einstein's path to becoming an influential physicist. The supportive environment of the Winteler home, combined with the academic rigor of the cantonal school, provided Einstein with both the emotional grounding and the intellectual stimulation needed to embark on his future groundbreaking scientific endeavors.

The Zurich Polytechnic: Love and Learning

In the autumn of 1896, at the age of seventeen, Albert Einstein enrolled at the Zurich Polytechnic, having successfully completed his entrance examinations on his second attempt after a preparatory year at the cantonal school in Aarau. The Polytechnic was known for its demanding standards and had a reputation as a nurturing ground for future scientists and engineers. Einstein chose to study in the section of the school devoted to teaching physics and mathematics to prospective teachers, a decision that aligned with his deep-seated passion for both subjects and his desire to secure a stable future career.

The academic environment at Zurich Polytechnic was rigorous. Courses were structured to provide a comprehensive foundation in theoretical and applied physics, along with advanced mathematics. Einstein, known for his independent nature, often found the traditional lecture-based teaching methods constraining. He preferred to study the subjects at his own pace, which occasionally put him at odds with some of his professors. Despite these differences, he excelled in physics and mathematics, showing a clear aptitude for abstract and conceptual thinking that would later define his scientific career.

During his formative years at the Zurich Polytechnic, Albert Einstein encountered Mileva Marić, a connection that would profoundly influence both his personal life and his early scientific endeavors. As the only woman in their academic division, Mileva stood out not only for her rarity but also for her intellectual capability. Coming from Serbia, she brought with her a distinctive perspective and a shared hatred for the traditional, often rigid educational methods that emphasized memorization over understanding.

Einstein and Marić found common ground in their passion for physics and mathematics, fields in which both showed exceptional promise. This mutual interest in the sciences became the foundation of their relationship, which evolved from a university friendship to a deep romantic involvement. Unlike the typical relationships of their time, their bond was significantly characterized by intellectual partnership. They engaged in lengthy discussions about complex scientific concepts, challenged each other's ideas, and debated vigorously on various academic principles. This ongoing intellectual engagement was an important element of their relationship, providing both stimulation and a shared purpose that deepened their connection.

The relationship between Einstein and Marić was not only personal but also collaborative. They spent hours together in study and discussion, which likely played a role in shaping some of Einstein's early thoughts and theories. Although the extent of Marić's contribution to Einstein's early work, including the theory of relativity, remains a subject of historical debate, it is clear that she was a significant intellectual companion during the critical years of his early career. Their correspondence often included discussions of their studies and speculations about new physics concepts, suggesting a partnership that was both personally and intellectually fulfilling.

Despite the strengths of their partnership, Einstein and Marić faced numerous challenges. Being a woman in a predominantly male field, Marić encountered significant obstacles, from gender bias in academia to the difficulties of balancing her ambitions with societal expectations. Einstein's support during these times was crucial, offering not just companionship but also encouragement against the discouraging academic climate for women.

The couple's collaboration is most notably highlighted during their final year at the Polytechnic when they tackled their graduation theses. Einstein's thesis on the dimensions of molecules provided insights that later contributed to the development of his groundbreaking theories. Although Mileva faced her own struggles, failing to graduate on her first attempt, her influence on Einstein's early scientific development was significant. The discussions and mutual critique of their academic work contributed to refining and challenging Einstein's ideas, pushing him to consider different perspectives.

Einstein's years at the Zurich Polytechnic were marked by intense learning, personal growth, and the forging of relationships that would influence both his personal and professional life. During this time there was a blend of academic challenges and romantic entanglements, set against the backdrop of the burgeoning field of theoretical physics*. The experiences and relationships formed during these years were instrumental in shaping his path to becoming one of the most iconic scientists of the twentieth century, paving the way for his later contributions to our understanding of the universe.

---

* Theoretical physics is a branch of physics that employs mathematical models and abstractions to explain and predict natural phenomena. It focuses on developing theories that can explain the fundamental components of the universe and the forces that govern their interactions.

The Patent Office and the Annus Mirabilis

After graduating in 1900, Albert Einstein was ready to embark on a career as a physics teacher. He was equipped with a diploma and deeply in love, seemingly ready for a promising phase of life. However, he faced significant problems and challenges: job opportunities were limited, and financial support from his uncle had stopped. With limited means, his clothing became worn, and meals were infrequent which adversely affected his health. These challenges also meant he could not afford to marry Mileva Marić, his companion and confidante.

In 1901, a year after his initial struggles, Albert Einstein successfully obtained Swiss citizenship, marking a significant milestone in his life. This new status as a Swiss citizen was integral for his future, offering him opportunities and rights within Switzerland. Despite this advancement, he was not required to fulfill military service obligations, which were typically mandatory for Swiss citizens. The exemption was granted by Swiss authorities due to medical reasons, although specific details about these health concerns are not extensively documented.

After graduation, Einstein faced significant challenges in securing a teaching position. Despite nearly two years of relentless job applications, Swiss schools consistently overlooked his potential, offering him no opportunities. The situation worsened when his father passed away, leaving Albert heartbroken. During this difficult time, Mileva's support was invaluable. She kept by his side, no matter how hard the financial situation became. Her unwavering dedication provided a source of strength and comfort, allowing him to focus on his work even when the odds seemed insurmountable.

The turning point in Einstein's early career came with the assistance of Marcel Grossmann's* father, who helped him secure a position at the Swiss Patent Office in Bern. The job at the patent office was not the academic or educational career he had envisioned, but it provided a necessary livelihood. Surprisingly, Albert's role at the Patent Office proved to be more rewarding than he initially anticipated.

At the Patent Office, Albert's task was to review and analyze the potential of inventions submitted for patents. This work required a sharp, analytical mind capable of understanding concepts and innovations in their early stages. Einstein excelled at this, often completing his daily responsibilities well ahead of schedule. This efficiency afforded him time to pursue his true passion – scientific inquiry and thinking.

With stable employment at the Swiss Patent Office, Albert Einstein found the security he needed to propose to Mileva Marić. She accepted and they married in 1903, marking the beginning of a profound partnership both personally and intellectually. The following year, their joy was compounded with the birth of their son, Hans Albert, enriching Einstein's family life further.

---

* Marcel Grossmann was a Swiss mathematician and a friend and classmate of Albert Einstein. Grossmann was a member of an old Swiss family from Zürich.

However, the publication of letters between Albert and Mileva in 1987 revealed a more complex layer to their early years together. These letters brought to light that in early 1902, while visiting her family in Novi Sad, Mileva gave birth to a daughter named Lieserl. The circumstances surrounding Lieserl's fate are poignant and mysterious; when Mileva returned to Switzerland, she did so without Lieserl. A letter from Einstein dated September 1903 suggests that the child might have been adopted or tragically died of scarlet fever during infancy, a loss that would have deeply impacted the couple.

Despite the shadows cast by this personal sorrow, Einstein's professional life provided a stable foundation for his burgeoning family. His role at the Patent Office was not just a job but a stepping stone that allowed him the mental space and time to pursue his scientific inquiries. The intellectual freedom and flexible time afforded by his job allowed him to focus deeply on complex problems, leading to an explosive period of creativity and productivity. In what would later be known as his "Annus Mirabilis," or miraculous year, 1905, Einstein published five seminal papers in the journal "Annalen der Physik," forever altering the landscape of modern physics.

This period was so productive that Einstein himself described it as one where "a storm broke loose in my mind," as this period became a testament to the profound insights and theoretical breakthroughs that characterized his later work. Among these papers was his doctoral dissertation, "Eine neue Bestimmung der Moleküldimensionen" (A New Determination of Molecular Dimensions), submitted on April 30, 1905, and later approved by Professor Alfred Kleiner at the University of Zurich. This work, dedicated to his friend Marcel Grossmann, was a significant contribution to molecular physics* and highlighted his deep analytical abilities. By July 1905, this dissertation

* Molecular physics is the branch of physics that studies the physical properties of molecules, the dynamics of their interactions, and the ways they behave under different conditions. This field explores how molecules form, how they react with one another, and how they absorb and emit radiation. Essentially, it involves understanding the physical basis of the chemical properties of molecules, which can be crucial for fields like chemistry, materials science, and biochemistry.

was instrumental in earning him a PhD from the University of Zurich, formalized on January 15, 1906.

Albert Einstein's first published paper, titled "Folgerungen aus den Capillaritätserscheinungen" (Conclusions Drawn from the Phenomena of Capillarity), appeared in 1900. This work, which Einstein himself later deemed relatively minor, actually marked the beginning of his impactful scientific publications. The paper focused on capillary forces, exploring how liquids rise or fall in thin tubes, which are smaller than the liquid's own dimensions, due to the interactions between the liquid and the surrounding surfaces. This early exploration of physical forces set the stage for Einstein's future contributions to the scientific community.

In one of his influential papers during this period, Albert Einstein delved into the photoelectric effect, proposing a revolutionary idea that light could be absorbed in distinct packets of energy, termed 'photons.' This concept was radical as it challenged the classical wave theory of light, which treated light purely as a wave phenomenon. Einstein's photon hypothesis suggested that the energy of these photons is directly related to the frequency of the light, not its intensity. This insight laid the foundational principles of quantum mechanics* and helped explain how materials absorb light and release electrons – a process that is still critical to the development of various technologies such as solar panels and photoelectric sensors.

Additionally, Einstein explored the phenomenon of Brownian motion, which involves the random movements of particles suspended in a fluid. Through statistical analysis, he provided empirical evidence supporting the atomic theory, which says that all matter is made up of atoms and molecules. Before his work, atoms were mostly a theoretical

---

* Quantum mechanics is a fundamental theory in physics that describes the physical properties of nature at the scale of atoms and subatomic particles. It introduces concepts that are quite different from classical physics, as it suggests that energy, momentum, and other quantities of a bound system are restricted to discrete values (quantization), objects have characteristics of both particles and waves (wave-particle duality), and there are limits to how precisely the values of some pairs of linked physical properties, such as position and momentum, can be known simultaneously (the uncertainty principle).

construct without direct empirical evidence. Einstein's analysis demonstrated that the visible, erratic movements of particles in a fluid were the result of collisions with atoms and molecules, thus offering a macroscopic proof of atomic theory. This work not only reinforced the concept of atoms but also played a crucial role in the acceptance of theoretical models of matter at the microscopic level.

In the same time, Einstein's groundbreaking paper on the special theory of relativity fundamentally transformed the field of physics by introducing a new framework in which the laws of physics are seen as consistent across all non-accelerating frames, meaning they hold true regardless of how these frames are moving relative to one another as long as they are not accelerating. A pivotal aspect of this theory is the constancy of the speed of light in a vacuum, which Einstein proposed remains the same no matter the speed at which an observer travels relative to the light source. This idea dramatically altered traditional concepts of time and space, which were previously viewed as absolute. Einstein suggested that time can dilate, or slow down, and lengths can contract depending on the relative speeds of the observers and objects in motion. This theory re-defined our understanding of space and time as interwoven continuums, reshaping how we perceive movement and velocity in the universe.

Building on the revolutionary ideas of his special theory of relativity, Einstein introduced the concept of mass-energy equivalence, encapsulated in the ionic equation $E=mc^2$. This equation asserts that mass and energy are interchangeable; it tells us that the mass of a particle can be converted into energy, and vice versa. Here, (E) represents energy, (m) represents mass, and (c) is the speed of light in a vacuum, squared. This relationship implies that a small amount of mass can be converted into a large amount of energy, illustrating a profound interconnectedness between these two fundamental properties. This concept has not only become one of the most famous equations in physics but also has practical implications, explaining the vast energy released in nuclear reactions and the functioning of stars, including our Sun. Through these ideas, Einstein not only advanced theoretical knowledge but also set the stage for advancements in energy production and our understanding of the universe's structure.

In addition to his scientific achievements, Einstein found intellectual camaraderie and stimulation in a discussion group he formed with friends in Bern, known as the Olympia Academy. This informal discussion group, named with a touch of irony to signify its modest nature compared to the grandiosity of the actual Olympic Games, was more than just a gathering of like-minded individuals; it was a vessel for the exchange and evolution of groundbreaking ideas.

The group met regularly in Einstein's apartment, which he shared with his wife, Mileva Marić. Mileva, herself a physicist, often participated in these meetings, even though she usually listened silently in comparison to the other members. Her contributions, though not as vocal, were rooted in a deep understanding of the subjects at hand, providing insightful observations and critical feedback on discussed topics.

The discussions at the Olympia Academy covered a wide range of subjects but predominantly focused on physics, mathematics, and philosophy. Members of the group, including Maurice Solovine and Conrad Habicht, were instrumental in challenging each other's ideas, fostering a rigorous intellectual environment that prompted Einstein to refine his thoughts and theories.

They all gave significant attention to the works of people like Henri Poincaré, Ernst Mach, and David Hume, each of whom had a profound influence on Einstein's thinking. Poincaré's work on the foundations of geometry and his philosophical questions about science and hypothesis testing prompted Einstein to think more deeply about the nature of space and time. Ernst Mach's principles, especially his ideas about the economy of science and skepticism of absolute space and time, resonated with Einstein and were instrumental in shaping his approach to the special theory of relativity. Additionally, David Hume's empiricism* and his rigorous questioning of causality and perception challenged Einstein to consider the assumptions underlying classical mechanics.

The discussions that took place within the walls of the Olympia Academy were paramount in shaping not only Einstein's scientific outlook but also his philosophical views. The intense debates and the freedom to critique existing theories without restraint encouraged a culture of intellectual rigor and open inquiry. This environment was significant for Einstein as it allowed him to question orthodox doctrines and norms, leading him to develop revolutionary theories that would later change the course of modern physics.

In this nurturing environment, Einstein thrived, balancing his roles as a scientist and a family man. This period was crucial, not only for his personal happiness but also for laying the foundational ideas that would revolutionize the understanding of the universe.

---

* Empiricism is a philosophical theory that emphasizes the role of experience and sensory perception in the formation of ideas, while discounting the notion of innate ideas. In simpler terms, empiricism posits that knowledge comes primarily from sensory experience. According to this view, all concepts and knowledge that humans possess arise from observing the world through the five senses.

Marriage and Turmoil: Personal Struggles Amidst Academic Triumphs

In January 1906, shortly after his remarkable annus mirabilis, Einstein was officially awarded a doctorate by the University of Zurich. His intellectual pursuits were rapidly gaining recognition, but his professional life at the Swiss Patent Office remained relatively simple and boring. By April, he received a promotion to a second-class technical expert, a position that allowed him some financial stability, though it was far from fulfilling his intellectual ambitions.

1907 marked a pivotal year for Einstein as he began to extend his theories beyond the confines of the work he had previously done. During this paramount period, Albert Einstein began to delve deeper into the complexities of gravitational theory, which laid the groundwork for his General Theory of Relativity. This innovative idea fundamentally changed our conception of the universe, suggesting that massive objects like planets and stars warp the fabric of spacetime around them, influencing the paths that other objects take through space.

Moreover, in 1907, building on his revolutionary insights, Einstein introduced the principle of equivalence, a key concept that extended the implications of his famous equation, $E=mc^2$, from his 1905 paper on the special theory of relativity. This principle states that the effects of gravity are indistinguishable from the effects of acceleration. For example, being in a closed box on Earth's surface where gravity pulls you down is indistinguishable from being in a closed box in space that is accelerating upwards. This concept was important because it bridged the gap between his earlier work on special relativity, which deals with systems moving at constant speeds, and his later work on general relativity, which addresses systems undergoing acceleration.

This insight was pivotal because it challenged the traditional Newtonian mechanics, which treated gravitational and inertial mass* differently. In Newtonian physics, inertial mass determines an object's resistance to acceleration, while gravitational mass determines the strength of gravitational attraction between masses. Einstein's equivalence principle suggested that these two types of mass were actually the same, thereby simplifying the conceptual framework of physics and laying the groundwork for his later development of the general theory of relativity.

Einstein's early assertion that mass and energy are interchangeable, encapsulated in the equation $E=mc^2$, indicated that mass can be converted into energy and vice versa. This idea not only supported the principle of equivalence but also led to the understanding that the energy content of a body is a measure of its mass – profoundly linking the previously separate concepts of mass (a measure of an object's amount of matter) and energy (the capacity to do work). This

---

* Inertial mass is a measure of an object's resistance to changes in its state of motion when a force is applied. It's essentially a property that quantifies how much an object will resist being accelerated by a force. In simpler terms, the greater the inertial mass of an object, the harder it is to change its velocity – whether that means speeding it up, slowing it down, or changing its direction.

connection has profound implications across physics, including providing the theoretical foundation for nuclear energy and helping to explain stellar processes in astrophysics.

This advancement was crucial not only for theoretical physics but also for our understanding of the universe at large. It allowed later physicists to make predictions about phenomena such as black holes, gravitational waves, and the behavior of objects in intense gravitational fields, such as those near stars and planets.

Despite these theoretical advances, not all of Einstein's academic endeavors were met with immediate success. His application for a doctorate at the University of Bern was initially rejected, as his submission was considered insufficient. Undeterred, Einstein submitted a new dissertation the following year, leading to his award of the doctorate at the University of Bern in 1908. This academic recognition allowed him to transition to an academic career, becoming a Privatdozent, or private college lecturer. By the end of 1908, he delivered his first lecture, marking his formal entry into the academic community.

The year 1909 marked a significant turning point in Albert Einstein's career, transitioning him from a competent patent clerk to a recognized figure in the academic world of physics. This transformation began in July, when Einstein received his first honorary doctorate from the University of Geneva. This honor, acknowledging his substantial contributions to theoretical physics, was a profound affirmation of his work's impact – a recognition that underscored his potential and heralded the beginning of many such acknowledgments.

Einstein's transition away from the Swiss Patent Office in October of the same year marked the end of a phase of his life where he balanced his bureaucratic responsibilities with his behind-the-scenes dedication to physics. The Patent Office job had provided a stable income that supported his young family, but intellectually, it was far from fulfilling. His departure from this role was a significant leap, driven by his growing reputation and the desire to devote himself fully to his scientific pursuits.

This leap was immediately followed by a substantial professional advancement. Einstein was appointed as an associate professor of theoretical physics at the University of Zurich. This position was not just a job; it was a recognition of his talents and an opportunity to influence the next generation of physicists. It also allowed him more time and resources to focus on his research, which continued to push the boundaries of contemporary understanding of physics.

Einstein's integration into the academic community was further solidified when he became a member of the Society of Physikalische Gesellschaft Zurich (Physical Society of Zurich) on December 2, 1909. Joining this society was not merely a ceremonial inclusion; it placed him squarely within the network of Europe's leading scientists and thinkers. Membership in such a society provided him with a platform to share his ideas, receive feedback from peers, and stay informed about the latest developments in physics and related disciplines.

In 1910, Albert Einstein's life was a blend of professional ascendancy and personal complexities. His acceptance of a professorship at the German University in Prague marked a significant milestone in his career. This move not only shifted his academic base outside Switzerland for the first time but also expanded his influence across a broader geographical and intellectual landscape. The position at Prague was a prestigious one, offering him a platform in one of Central Europe's historically renowned academic centers. This period in Prague, though brief, was critical as it exposed him to new academic challenges and collaborations, enhancing his reputation in the international scientific community.

However, 1910 was also a year of personal challenges and changes. The birth of his second son, Eduard, on July 28, brought great joy to Einstein. Eduard, whom Einstein affectionately called 'Tete,' added a new dimension to his family life. The joy of Eduard's arrival, however, was juxtaposed with the growing strains in Einstein's marriage to Mileva Marić. The couple's relationship, once based on mutual intellectual admiration and collaboration, had begun to deteriorate under the weight of Einstein's career demands and Mileva's growing discontent with their domestic situation. Mileva, herself an accomplished physicist, had put aside her own career aspirations to focus on the family. It was a sacrifice that bred resentment as Einstein's career flourished and took him away from home frequently.

The personal strains were further compounded by Mileva's struggles with her own identity and career, feeling increasingly isolated in the shadow of her husband's rising prominence. As Einstein's focus on his work intensified, the emotional distance between them grew. This growing rift was, at times, tempered by their mutual commitment to their children and to maintaining a stable home life, but the underlying tensions were increasingly prominent.

Amid these personal challenges, Einstein's professional life continued to thrive. His election on November 14 as a member of the Naturforschende Gesellschaft of Zurich (Natural Science Society of Zurich) was a significant honor, symbolizing his deepened integration into the scientific community. This membership not only acknowledged his contributions to science but also provided a vital intellectual network. However, even as his professional circle widened, the personal troubles at home made 1910 a year of contradictions.

These years, filled with academic accolades and personal milestones, illustrate the complex weave of Einstein's life during this period. His professional life was soaring with awards and positions that recognized his theoretical genius, while his personal life saw him failing to balance roles as a father and a good husband in the middle of his burgeoning scientific career. Each step forward in his professional journey was paralleled by life events that anchored him in profound ways, reflecting the duality of his journey through the realms of theoretical physics and familial responsibilities.

Crossroads of Thought: From Zurich to Berlin

Albert Einstein's journey through the early 1910s marked a period of intense personal and professional development, a time when his theoretical insights began shaping the future of physics. In 1911, he stepped into a significant new role as a full professor at the German University in Prague. This position not only marked his climb within the academic community but also challenged him to deepen his theoretical explorations. Prague served as a fertile ground for Einstein's growing ambitions in physics. It was here that he began to seriously calculate the deflection of light by gravity, a phenomenon he predicted could be observed during a solar eclipse. This insight laid the groundwork for what would become a crucial experimental test of his developing theory of general relativity.

Einstein's time in Prague was also noteworthy for his participation in the first Solvay Congress in Brussels in October 1911. This congress gathered many of the leading scientific minds of the time, providing Einstein with a platform to discuss his ideas and theories with other eminent physicists. It was a paramount moment that not only elevated his stature among his peers but also sharpened his theoretical perspectives.

In 1912, as Albert Einstein's academic career was reaching new heights, his personal life also underwent a significant transformation. His relationship with Elsa Löwenthal, who was both his first cousin on his mother's side and his second cousin on his father's side, began to take a prominent place in his life. This relationship would later evolve into a cornerstone of his personal and emotional world.

Elsa, unlike Mileva, Einstein's first wife, was not a scientist, but she provided him with a different type of companionship, one that offered emotional support and domestic stability. Their connection rekindled during Einstein's professional visits to Berlin and quickly grew into a deep bond. This relationship was particularly significant, as it not only introduced a new chapter in Einstein's personal life but also eventually influenced his decision to move to Berlin, marking a critical junction in both his personal and professional paths.

Parallel to these developments in his personal life, 1912 was also a pivotal year for Einstein professionally. He accepted an invitation to return to Zurich, taking up a full professorship at the ETH Zurich, the very institution where he had previously studied and later taught. This return to Zurich represented a significant homecoming. It was a return to his academic roots, to a place that had been foundational in shaping his early career. Here, Einstein was greeted not just by familiar sights and old colleagues but by an intellectual environment that he had long cherished.

His position at ETH allowed him to focus intensively on his research without the heavy teaching responsibilities that typically accompany university roles. It was during this time that Einstein began to delve even deeper into the complexities of the theoretical framework that would become the general theory of relativity.

During this period, Einstein began collaborating with Marcel Grossmann, his friend and a professor of mathematics at the ETH. Grossmann's expertise in mathematics was instrumental for Einstein, particularly in grappling with the complex mathematical formulations required by his nascent theory of general relativity. This collaboration was highly important, as Grossmann introduced Einstein to tensor calculus, a type of mathematics involving transformations between different coordinate systems. Tensor calculus would become essential in the formulation of general relativity. It allows for the description of physical laws in a form that is independent of the choice of coordinates, making it ideal for dealing with the curved spacetime of general relativity where traditional coordinate systems may not apply straightforwardly. This mathematical framework enabled Einstein to precisely articulate how mass and energy influence the structure of space and time. Their work during these years was foundational, setting the stage for the theoretical breakthroughs that would soon follow.

The year 1913 was a turning point for Einstein. His reputation as a leading theoretical physicist had grown significantly, catching the attention of two prominent figures in the German scientific establishment: Max Planck and Walther Nernst. They traveled to Zurich with a proposition that would dramatically alter Einstein's career trajectory. They offered him membership in the prestigious Prussian Academy of Sciences and a professorship at the University of Berlin. Crucially, this position did not require him to teach, allowing him maximum freedom to focus on his research. They also proposed that he manage the yet-to-be-founded Kaiser Wilhelm Institute for Physics.

The offer was appealing for several reasons, one of them was that Berlin was already home to Elsa, the woman Einstein was having an affair with. Beyond personal reasons, the position represented an unparalleled opportunity for Einstein to advance his research in a vibrant academic environment that was at the forefront of scientific innovation.

On November 12, 1913, Wilhelm II, German Emperor and King of Prussia, officially approved Einstein's membership in the Academy. By December 7, Einstein formally accepted the offer from Berlin, setting the stage for the next significant phase of his career and life.

The years from 1911 to 1914 were a period of profound growth and change for Einstein. They were years filled with new challenges, opportunities, and personal developments. Professionally, Einstein moved through positions that offered increasing levels of freedom and responsibility, allowing him to refine his theories that would soon revolutionize our understanding of the universe. Personally, his life was becoming more intertwined with Elsa's, who would soon become a major figure in his life.

As Einstein prepared to leave Zurich for Berlin, he was not merely moving to a new city; he was stepping into a future that promised even greater opportunities for theoretical exploration and personal fulfillment. This transition was not just a geographical move but a significant leap into a new chapter that would see the birth of his general theory of relativity and his emergence as a scientist of international renown.

Berlin and Elsa: A New Chapter in Life and Love

In 1914, as the world stood on the brink of monumental change with the onset of World War I, Albert Einstein's life was undergoing its own significant transformations. Arriving in Berlin at the beginning of April, he was ready to begin a new chapter both personally and professionally, at a time when the city itself was about to navigate the turbulent waters of global conflict and scientific revolution.

Upon his arrival, Einstein was joined by his wife Mileva and their two sons, Hans Albert and Eduard, who arrived a month later. This reunion was, however, short-lived. By July of that year, the personal difficulties had reached a tipping point, and Mileva decided to move back to Zurich with their sons. This decision was not made lightly; it reflected deep unresolved issues in their marriage, including Einstein's affair with Elsa and Mileva's dissatisfaction with their life in Berlin. The move was a physical manifestation of the widening gulf between

Einstein's personal predicaments and his professional life. Mileva's return to Zurich symbolized a retreat to a familiar environment where she had her own support network and could feel more autonomous.

Einstein's introductory address at the Prussian Academy of Sciences was a landmark moment in his career in the same year, signaling his acceptance into the upper echelons of European science. This achievement, however, came at a time when he was losing his family. The event not only highlighted his professional ascent but also marked the beginning of the end of his marriage. The prestige of joining such a distinguished institution contrasted sharply with the unraveling of his personal life.

As World War I erupted across Europe, it brought with it a surge of nationalistic fervor and militaristic passion that deeply troubled Albert Einstein. He observed with increasing alarm how the enthusiastic war spirit took society and was often unquestioned by the intellectual community. Disturbed by these developments, Einstein's worldview began to incorporate a more active and outspoken stance on broader societal issues, particularly peace and disarmament.

During this turbulent period, Einstein aligned himself with the Bund Neues Vaterland, a newly formed pacifist organization that aimed to counteract the prevailing pro-war sentiments and promote peace. The group consisted of intellectuals, artists, and academics who shared a common hatred for the war and the nationalistic ideologies fueling it. This association was not just a passive membership for Einstein; it signified a profound shift toward active political engagement, reflecting his ideological move from detached scientist to public intellectual advocating for peace.

Einstein's involvement in the pacifist movement was further stressed by his endorsement of the "Manifesto to the Europeans," penned by his friend and fellow pacifist Georg Nicolai. This manifesto was a direct response to the widespread "Manifesto of the Ninety-Three," which had been signed by many of Germany's leading scientists and intellectuals, justifying Germany's military actions and war policies. In stark contrast, the "Manifesto to the Europeans" called for rational thinking and a unified, peaceful Europe free from the divisive and destructive forces of nationalism.

Einstein's decision to sign this counter-manifesto was emblematic of his deep-seated belief in internationalism and his rejection of the tribal loyalties that he saw as a root cause of the conflict.

Einstein's public stance on these issues was not without consequences. It isolated him from many of his peers who supported the war and viewed his pacifism as unpatriotic, if not outright traitorous. This period of conflict and engagement sharpened Einstein's resolve to use his platform not just for scientific discourse but as a means of promoting peace and rational discourse in political matters.

This deepening engagement with political issues was a momentous development in Einstein's life and career. It marked the beginning of his lifelong commitment to social and political causes, ranging from pacifism and disarmament to civil rights and Zionism*. His advocacy for peaceful solutions and his critique of militarism during World War I laid the groundwork for his later involvement in international politics, including his support for the establishment of the League of Nations and his efforts in the anti-Nazi movement.

In 1915, Albert Einstein delved deep into experimental physics through his collaboration with J.W. de Haas, a Dutch physicist known for his work in magnetism and experimental physics, focusing on what became known as the Einstein-de Haas effect. This phenomenon demonstrates the fundamental relationship between magnetism and angular momentum, two important concepts in physics.

The gyromagnetic effect, as studied by Einstein and de Haas, essentially explored how a magnetic field can influence the motion of electrons within a material, leading to measurable mechanical

---

* Zionism is a nationalist and political movement that supports the establishment and maintenance of a Jewish state in the territory defined as the historic Land of Israel (roughly corresponding to Canaan, the Holy Land, or the region of Palestine). The movement originated in the late 19th century in response to growing anti-Semitism across Europe, with Theodor Herzl, a key figure in its establishment, promoting the idea that the best way to avoid discrimination against Jews was to have their own nation.

rotation. Its working revolved around the experiment that when a ferromagnetic material – typically iron – is magnetized, it experiences a change in momentum due to the alignment of the magnetic moments (or spins) of electrons. According to the laws of conservation of angular momentum, this change inside the material had to be balanced by a corresponding physical rotation of the material itself if it was free to move.

This experiment confirmed the direct link between the microscopic quantum world of electron spin and the macroscopic classical world of mechanical rotation. By demonstrating that magnetizing a material could cause it to spin, Einstein and de Haas bridged the gap between these two realms, providing empirical support for the theory of angular momentum conservation in electromagnetism. This work not only deepened the understanding of fundamental physics concepts but also reinforced the quantum theory that was still in its formative years at the time.

But it was his work on the general theory of relativity that would become the year's hallmark. In November, after intensive theoretical development, Einstein presented this groundbreaking theory in a four-part lecture series at the Prussian Academy of Sciences. The general theory of relativity proposed that gravity is not a conventional force acting at a distance. Instead, Einstein suggested that gravity is a result of the curvature of spacetime, which is the four-dimensional fabric that combines the dimensions of space and time. According to this theory, massive objects like planets and stars warp the spacetime around them, and this curvature dictates how objects move in space.

In essence, when we see a planet orbiting a star or an apple falling from a tree, these objects are following paths determined by the curved spacetime created by the mass of the star or the Earth. This idea was a radical departure from the notion that objects influence each other through invisible forces at a distance.

Einstein's theory also introduced the concept that mass and energy could affect the curvature of spacetime. This interconnection between mass, energy, and the geometry of space forms the core of how general relativity describes the universe. This groundbreaking shift in understanding required a new mathematical framework, which Einstein developed from the mathematics of metrics and tensors, explaining how the presence of mass and energy "tells" spacetime how to curve, and spacetime "tells" objects how to move. This profound relationship between spacetime and the objects within it redefined not only the study of gravity but also laid the foundational concepts for modern cosmology and astrophysics.

Owing to this hallmark theory, On December 18, 1915, Einstein's contributions were further recognized when he was elected as a corresponding member of the Royal Society of Göttingen.

The year 1916 saw the publication of "Die Grundlage der allgemeinen Relativitätstheorie" (The Formal Foundation of the General Theory of Relativity) in the Annalen der Physik, formalizing his theories for the scientific community. That year, Einstein also succeeded Max Planck as the head of the Deutsche Physikalische Gesellschaft (German Physical Society). In December, he completed "Über die spezielle und die allgemeine Relativitätstheorie, gemeinverständlich" (On the Special and General Theory of Relativity, A Popular Account), making his complex theories more accessible to the public.

The year 1917 came with problems as Einstein's health declined. He started suffering from a liver ailment and a stomach ulcer, among other issues. During this time of physical vulnerability, his cousin and now romantic partner, Elsa, took on the role of caregiver, nurturing him back to health. Their relationship deepened, setting the stage for their eventual marriage. Despite his health challenges, Einstein did not cease his intellectual pursuits; he explored cosmological ideas, introducing the cosmological constant in an attempt to achieve a stable universe model, a concept he would later famously renounce as his "biggest blunder."

By 1918, the war's end brought political upheavals that Einstein welcomed, though he declined a joint offer from the Swiss Polytechnic and the University of Zurich, choosing to stay in Berlin. His personal life took a joyful turn in 1919, when, following his divorce from Mileva in February, he married Elsa on June 2. That same year, the observational confirmation of the deflection of light by the Sun's gravity during the solar eclipse of May 29 – predicted by his general theory – propelled Einstein to international fame.

Einstein's general theory of relativity proposed that what we perceive as the force of gravity actually arises from the bending of space and time by mass and energy. Einstein suggested that massive objects like stars and planets warp the fabric of space-time around them, and that other objects move along these bends, which is perceived as gravitational attraction.

In simpler terms, Einstein's theory posited that space and time are interwoven into a single continuum known as space-time. Massive objects cause a distortion in this space-time, which is akin to placing a heavy ball on a stretched rubber sheet. Smaller balls placed near the heavy ball will roll toward it not directly because of an active pull by the heavy ball, but because of the way it curves the sheet around it.

One of the intriguing predictions of general relativity was that light itself would be deflected when passing near a massive object. This effect occurs because the mass of the object warps the space-time through which the light travels, effectively bending the path of the light. The amount of bending predicted by Einstein could only be tested during a solar eclipse, when the brightness of the Sun is blocked enough that stars close to the Sun in the sky can be observed.

The confirmation came during the solar eclipse of May 29, 1919, when observations made by British astronomers Arthur Eddington and Frank Watson Dyson in two different locations (Principe Island off the coast of West Africa and Sobral in Brazil) demonstrated that light from stars passing close to the Sun was indeed bent. The amount of bending was consistent with the predictions made by general relativity, and not with the predictions of Newtonian physics.

The successful confirmation of this prediction was a stunning triumph for Einstein's theory. It not only validated his ideas about the nature of gravity and the structure of the universe but also transformed him into a global scientific celebrity almost overnight. His breakthroughs in physics were not just discussed within the selective confines of scientific institutions but became a topic of fascination for the general public worldwide.

Newspapers around the world reported on the results of the eclipse expedition, often in sensational terms, pointing out Einstein as the successor to Isaac Newton. Thus, the "myth of Albert Einstein" began to take shape, framed both by his scientific brilliance and his emerging status as a public intellectual. This public recognition transformed Einstein into "perhaps the world's first celebrity scientist," a figure who had revolutionized a framework that had underpinned physicists' understanding of the universe since the seventeenth century.

This newfound celebrity status marked a significant transformation in Einstein's career and public persona. His theory of general relativity, once obscure and known only to a handful of physicists, now challenged long-held beliefs about the universe, placing Einstein at the forefront of a scientific revolution. His rise to fame was propelled not only by the novelty and wisdom of his ideas but also by a growing public curiosity about science and its potential to unravel the mysteries of the universe.

Einstein's interactions with Kurt Blumenfeld, a prominent leader in the Zionist movement, were also particularly influential in shaping his political activities throughout the late 1910s. Blumenfeld introduced Einstein to the ideals of Zionism, which aimed at establishing a homeland for the Jewish people. Einstein found the cultural and humanitarian aspects of Zionism deeply compelling. He saw it as a movement that could foster Jewish education and culture, which resonated with his own ideals about freedom and individual rights. His discussions with Blumenfeld encouraged him to use his international stature to advocate for Zionist causes, a role he embraced throughout the 1920s and beyond. This engagement reflected a significant expansion of Einstein's public persona, from a physicist to an advocate for Jewish communal development.

1919 was also a pivotal year for Einstein's personal life. His marriage to Elsa Löwenthal in June brought new changes, as he welcomed Elsa and her two daughters, Ilse and Margot, into his life. Ilse was born in 1897 and Margot in 1899. Therefore, at the time of their mother's marriage to Einstein in 1919, Ilse was about 22 years old and Margot about 20 years old. This new family dynamic provided him with a stable home life and emotional support, which was crucial during a period of intense public scrutiny and professional demand. Elsa played an instrumental role as a caretaker and manager, handling the practical aspects of their lives, which allowed Einstein to devote more time to his scientific and political pursuits. The integration of his family life with his increasing public role also influenced how he managed his time and commitments, balancing his domestic responsibilities with his commitments to science and public advocacy.

The recognition of Einstein's contributions to science continued to grow in 1919, reaching their heights when he received an honorary doctorate from the University of Rostock. This honor was not just a recognition of his theoretical achievements but also a testament to his rising status as a leading figure in the scientific community. The award from the University of Rostock was particularly significant, coming at a time when Einstein's theories were beginning to gain widespread acceptance and his public profile was expanding beyond the academic world.

These years from 1914 to 1919 were transformative for Einstein, marked by intense personal changes and profound professional achievements. As he navigated his new life in Berlin, his theories not only reshaped the foundations of physics but also began to redefine his role on the world stage, setting the groundwork for his enduring legacy as both a scientist and a thinker.

We're half way there: A note from Scott Matthews

As we find ourselves at the midpoint of the book, delving deeper into the profound journey of Albert Einstein's life, I want to express my sincere gratitude to you. The crafting of this narrative has been a labor of dedication, fueled by a profound reverence for Einstein's genius and a commitment to sharing his story with you.

Your engagement and reflections are invaluable not only to me but also to those who seek to comprehend the significance of Einstein's legacy. Your reviews not only support my endeavor as a storyteller but also contribute to the collective appreciation of Einstein's contributions to science and humanity. I take to heart each review, treasuring your perspectives and suggestions for further exploration.

If you have found resonance in the pages thus far or have ideas on how we can navigate the remainder of Einstein's journey together, I encourage you to take a moment to share your thoughts. A QR code is provided below for your convenience, which takes you to Amazon where you can leave your review. Whether you are reading digitally or holding a physical copy, a simple scan or click allows you to contribute your reflections.

Thank you for joining in this odyssey. Your feedback not only shapes the narrative but also honors Albert Einstein's enduring spirit. Here's to the unfolding chapters and the profound lessons they hold.

Theory, Recognition, and Responsibility

From 1919 to 1922, Albert Einstein's life was characterized by profound intellectual achievements and escalating public recognition, which brought with it both accolades and controversies. This period saw Einstein grappling with the implications of his theories, facing ideological opposition, and engaging with global political issues, marking a significant evolution in his role as a scientist and a public figure.

The year 1919 began on an extremely sad note with the death of Einstein's mother, Pauline, in February, after a severe illness in Berlin. Her death came at a time when Einstein was about to receive worldwide acclaim. The confirmation of his general theory of relativity during the solar eclipse of May 29, 1919, had made him an international celebrity.

His newfound fame was not without its challenges. As Einstein's theories began to gain popularity, they also attracted ideological and anti-Semitic* attacks, particularly in Germany, where some critics dismissed relativity as "Jewish physics." Despite this, Einstein's academic recognition continued to grow. He became a foreign member of the Royal Danish Academy of Sciences and Letters in April and met with Niels Bohr, marking the beginning of an impactful yet sometimes controversial collaboration. Niels Bohr was a prominent Danish physicist who made foundational contributions to understanding atomic structure and quantum theory. This collaboration between Einstein and Bohr was significant because it brought together two of the most influential figures in 20th-century physics. In May, he was also elected a foreign member of the Royal Dutch Academy of Sciences.

By 1919, Einstein's theories had begun to challenge the traditional framework that had dominated physics for more than two centuries. His assertion that space and time were dynamic and interwoven into a fabric that could be warped by mass was revolutionary and, for many, quite radical. These ideas not only redefined concepts of gravity but also implied a universe far more strange and changeable than previously thought.

The public debate surrounding Albert Einstein's theories of relativity, particularly his general theory about gravity, reached a critical moment on August 24, when a controversial presentation was held at the Berlin Philharmonic. Einstein, who attended the presentation, was well aware of the stakes. During that evening, the criticism directed at him varied widely. Some critics were skeptical and questioned the mathematical formulations he presented, while others outright dismissed the fundamental concepts of his theories. Following the event, Einstein chose to respond not in the lecture hall but through the press. His response, published in the Berliner Tageblatt, was not

---

* Anti-Semitic refers to hostility, prejudice, or discrimination against Jewish people. This term encompasses a range of negative feelings, actions, or expressions directed toward Jews simply because of their religious beliefs or ethnicity. It can manifest in various forms, including social exclusion, verbal attacks, written propaganda, and physical violence.

merely a defense of his scientific work but also an articulation of his stance on the role of science in society.

In his rebuttal, Einstein addressed the scientific criticisms head-on, clarifying misconceptions about his theories and providing detailed explanations to counter the claims of his critics. More importantly, he asserted the significance of scientific inquiry free from ideological constraints, emphasizing that the pursuit of knowledge should not be hindered by the personal biases or cultural prejudices of the observer.

The controversy at the Berlin Philharmonic thus became a defining moment in the early reception of relativity theory. It exemplified the tumultuous path of scientific progress and the personal resilience required of those who dare to redefine the boundaries of human knowledge.

Among the scientific criticism was also the growing anti-Semitism in Germany. Amid this growing anti-Semitic and academic hostility, Einstein considered leaving the country. His international engagements, however, continued to expand. He was invited to hold an introductory address as a guest professor at the University of Leiden in the Netherlands in October 1920, strengthening his ties with leading global scientific communities.

In 1921, Albert Einstein's first trip to the United States was marked by a plan that blended his rising celebrity status with his advocacy for the Zionist cause, alongside the renowned Zionist leader Chaim Weizmann. This tour was significant not only for Einstein's personal contributions to science and humanitarian efforts but also for the connections he established with major American institutions and figures.

One of the most notable stops on Einstein's tour was Princeton University, where he delivered four lectures on the theory of relativity. These lectures were later compiled and published, making them accessible to a broader audience beyond those who could attend in person. His visit to Princeton was highly publicized and helped solidify his fame in the American academic community. Einstein's arrival in New York was also met with considerable public and media attention. He attended various events organized by the local Jewish community and spoke on behalf of the Zionist cause, discussing the importance of establishing a Jewish homeland and the role of science and education in building societal foundations.

He also gave a series of lectures at Columbia and Harvard University, further discussing his theories and engaging with the academic community. These lectures were pivotal in spreading his scientific ideas among American scholars and students. In Washington, D.C., Einstein and Weizmann were received by President Warren G. Harding at the White House. This meeting was not only a significant acknowledgment of Einstein's growing influence but also a moment to discuss the potential for scientific and educational collaboration between the U.S. and the Jewish community in Palestine.

Beyond these major stops, Einstein's tour included public lectures and fundraising events for the Hebrew University of Jerusalem, which was then in the planning stages. His advocacy for the university underscored his commitment to education and his belief in the power of academic institutions to foster cultural and intellectual growth. By interacting with both the scientific community and the public, he not only disseminated his revolutionary theories but also drew attention to the Zionist movement, enhancing support for Jewish educational and cultural endeavors.

Afterward, in his 1921 essay, "My First Impression of the U.S.A.," Albert Einstein offered a thoughtful and positive perspective on American culture and society, drawing a sharp contrast between the vibrancy he observed in the United States and the prevailing atmosphere of exhaustion and disillusionment in post-war Europe. His reflections, much like those of Alexis de Tocqueville in the earlier era, provided an outsider's appreciation for the unique cultural dynamics of America, emphasizing traits that seemed to define the American spirit during the early 20th century. Alexis de Tocqueville was a French political thinker and historian best known for his work "Democracy in America." In the 1830s, Tocqueville traveled across the United States, and his observations and analyses of American society and political culture in this work have become foundational in the study of American and democratic societies.

Einstein's observations delved into the American optimism and the forward-looking attitude that permeated society. He was particularly struck by the general sense of well-being and the positive approach to life challenges, viewing these traits as fundamental to the American identity. He noted the friendly demeanor and self-confidence of the Americans he encountered, traits that appeared to be woven into the fabric of everyday life.

Additionally, Einstein's commentary shed light on the American capacity for innovation and the lack of envy, which he saw as enabling a more dynamic and fluid social and economic structure. This environment, as Einstein perceived, fostered greater opportunities for personal and collective advancement, contrasting sharply with the more rigid and hierarchical systems prevalent in many parts of Europe at the time. However, this did not diminish his engagement with European intellectuals or his influence across the continent.

Upon his return from the United States, Einstein visited Great Britain, where he delivered lectures in Manchester and London, further sharing his ideas and fostering international academic exchanges. His visit was not only a testimony to his growing global influence but also an opportunity to engage with the British scientific community. The University of Manchester recognized his contributions to science by awarding him an honorary doctorate, a significant honor that showcased the respect he commanded among his peers.

In 1922, Einstein's influence continued to expand as he delivered speeches in Prague and Vienna, cities that were then pivotal centers of intellectual thought in Europe. These engagements helped consolidate his reputation as one of the foremost physicists of his time, drawing attention to both his groundbreaking work in theoretical physics and his broader philosophical insights into science's role in society.

During the same period, Albert Einstein was also deeply engaged in further developing his theoretical framework; he submitted his inaugural paper on the unified field theory to the Academy of Science. This ambitious endeavor aimed to unify all the fundamental forces of nature – such as gravity, electromagnetism*, strong nuclear, and weak nuclear forces – into a single theoretical framework. This pursuit built upon his earlier monumental successes with the theories of relativity, which had already transformed our understanding of gravity and spacetime. The concept of a unified field theory represented a profound step in Einstein's lifelong quest to reveal a deeper, more interconnected understanding of the universe's structure. It showcased his relentless drive for scientific advancement and innovation, reflecting his belief that the complexities of the universe could be understood in a simpler, unified form. This work was momentous in setting the stage for future explorations in theoretical physics, paving the way for theories that seek to describe the universe in a comprehensive way.

In 1922, amidst his numerous scientific endeavors and global engagements, Albert Einstein undertook a significant diplomatic mission – a trip to France with the intent of fostering reconciliation between Germany and France in the aftermath of World War I. This visit was emblematic of Einstein's deepening involvement in international politics and his commitment to promoting peace and cooperation between former wartime adversaries. The visit to France was not just symbolic; it involved practical engagement with French scientists and intellectuals, public lectures, and discussions aimed at bridging the cultural and scientific divides exacerbated by the war. Einstein's efforts were part of a broader movement among intellectuals of the time who believed that science and cultural exchange could serve as conduits for peace.

---

* Electromagnetism is a branch of physics that explores the interaction between electrically charged particles. It combines the study of electricity, where charged particles produce electric fields and currents, and magnetism, where charges in motion generate magnetic fields. This fundamental force is responsible for practically all phenomena encountered in daily life above the nuclear scale, with the exception of gravity.

Einstein's presence in France, a country that had suffered immense losses during the war, was a powerful gesture of peace and mutual understanding. He spoke at prestigious institutions, including the Collège de France, advocating for greater scientific collaboration across borders and emphasizing the unifying potential of scientific inquiry. His visit also included poignant moments such as touring the battlefields of World War I, where he witnessed firsthand the scars left by the conflict on the landscape and the people. This experience further solidified his resolve to advocate for peace.

However, 1922 also brought significant personal challenges and public threats that underscored the volatile political climate in Germany. The assassination of Walther Rathenau, the German Foreign Secretary, was a stark reminder of the dangerous currents of extremism and nationalism that were gaining strength in Germany. Rathenau, a Jew like Einstein, was a prominent supporter of the Weimar Republic, which was the democratic government in Germany established after World War I. The Republic was founded in 1919, following the abdication of Kaiser Wilhelm II, marking the end of the German Empire. It was named after the city of Weimar, where its first constitutional assembly took place. Rathenau played a significant role in trying to stabilize Germany post-war, navigating the political and economic turmoil of the time. His assassination by right-wing extremists on June 24, 1922, shocked the nation and had a profound impact on Einstein.

Einstein's reaction to Rathenau's assassination was deeply personal and political. He saw Rathenau not only as a political figure but also as an ally in the struggle against militarism and extremism. In response to this tragedy, Einstein penned a poignant obituary, in which he grieved the rise of extremism in Germany and criticized the complicity of silence among the German population and officials who allowed such ideologies to flourish. His words were a mournful reflection on the loss of a friend and a warning about the dangerous path Germany was treading.

Due to increasing threats to his own safety, prompted by his outspoken political views and Jewish identity, Einstein found himself compelled to cancel all public appearances shortly after Rathenau's death. This period marked a significant shift in his life, as he became more cautious about his public engagements and deeply disillusioned with the political situation in Germany.

These events in 1922 illustrated Einstein's growing role as a public intellectual who did not shy away from engaging with political issues. His trip to France and his response to the political turmoil in Germany reflected his commitment to using his stature as a scientist to advocate for broader humanitarian and political causes, striving to promote peace in a troubled post-war Europe.

Despite the challenges faced throughout 1922, the year concluded positively for Einstein as he was awarded the Nobel Prize in Physics. This prestigious recognition was specifically for his explanation of the photoelectric effect, rather than for his theories of relativity, which were still considered controversial at the time. The photoelectric effect is a phenomenon in which electrons are emitted from materials, typically metals, when they absorb light of certain frequencies. Einstein's groundbreaking explanation proposed that light is composed of particles, or quanta, now called photons. He suggested that these photons carry energy proportional to their frequency, a radical idea that contradicted the classical wave theory of light but was crucial for the development of quantum mechanics. His theory provided a deeper understanding of the underlying mechanisms of the effect and supported the quantum theory that energy is exchanged only in discrete quantities. This insight not only solved a longstanding puzzle in physics but also laid the foundation for modern quantum physics, influencing the development of various technologies, including those used in digital cameras and solar cells.

Einstein on the World Stage: From Theory to Global Voice

After his groundbreaking theories had begun reshaping the understanding of the universe, Einstein continued to travel extensively, using his international fame to promote peace and intellectual cooperation.

In 1922, Albert Einstein embarked on an extensive six-month tour of Asia, speaking in various countries, including Japan, Singapore, and Sri Lanka (then known as Ceylon). His journey began in Tokyo, where after delivering his first public lecture, he was granted an audience with Emperor Yoshihito and his wife at the Imperial Palace. It was on this trip that he learned that he had been honored with a Nobel prize.

The event with the emperor drew massive crowds, with thousands of spectators lining the streets, eager to catch a glimpse of the famed physicist. Einstein observed that the Japanese people appeared modest, intelligent, and particularly appreciative of art, noting their strong artistic sensibilities in a letter to his sons. However, his private diary entries provide a less flattering view, questioning the intellectual interests of the nation compared to their artistic ones. Einstein writes: "Japanese are unostentatious, decent, altogether very appealing. Pure souls as nowhere else among people. One has to love and admire this country. However, the intellectual needs of this nation seem to be weaker than their artistic ones – natural disposition?"

Einstein's perceptions during his travels were not limited to Japan; he also recorded his impressions of China and India, though these were less complimentary. He remarked critically on the spirit and intelligence of the Chinese people, expressing concern about the implications of their broader influence on global diversity.

The culmination of his Asian expedition was his visit to Palestine, where he was greeted with exceptional enthusiasm. The British High Commissioner, Sir Herbert Samuel, welcomed him with honors typically reserved for heads of state, including a cannon salute. During a particularly crowded reception, Einstein expressed his pleasure at seeing Jews beginning to be recognized as a significant force worldwide.

In this journey to Palestine, he became the first honorary citizen of Tel Aviv*. This visit underlined his strong support for Zionist causes and his vision for the city as a future cultural and intellectual center for the Jewish community. Continuing his international engagements, Einstein then visited Spain, where he received a warm reception and was honored by the Royal Academy of Exact, Physical, and Natural Sciences in Madrid. He was awarded a corresponding membership and later an honorary doctorate, further recognizing his contributions to science and his growing influence in global intellectual circles.

---

* Tel Aviv, founded in 1909, is a vibrant city located on the Mediterranean coast of what was then part of the Ottoman Empire and later became part of British-administered Mandatory Palestine. It is known for its rich cultural heritage, modernist Bauhaus architecture, and as a center of economic and technological development. By the time of Albert Einstein's visit in 1923, Tel Aviv had evolved rapidly from a suburb of Jaffa into a significant urban center in its own right, embodying the aspirations of the Jewish community in Palestine for renewal and self-determination.

However, Albert Einstein's resignation from the League of Nations' Commission for Intellectual Cooperation in 1923 was a significant gesture that underscored his growing disillusionment with formal political institutions and their effectiveness in driving real, impactful change. Despite his initial enthusiasm for the Commission's potential to promote international intellectual collaboration and peace through dialogue and education, Einstein found the bureaucratic processes and political maneuverings within the League frustratingly counterproductive. He felt that these inefficiencies severely hindered the organization's ability to foster genuine intellectual cooperation and address the pressing issues of the time.

His decision to step down was rooted in a broader critique of how political agendas often overshadowed the substantive scientific and cultural exchanges that he valued highly. Einstein was deeply committed to the ideal of using science and education as bridges between conflicting nations and cultures, promoting understanding and cooperation. He believed in the transformative power of intellectual engagement to overcome nationalistic and ideological divides, a belief that was integral to his identity both as a scientist and a global citizen.

Despite stepping down from the League of Nations' Commission for Intellectual Cooperation in 1923, Albert Einstein remained deeply committed to fostering global intellectual exchanges. His dedication to international collaboration was particularly evident through his involvement with the "Freunde des Neuen Rußland" (Friends of New Russia), an organization established after the Russian Revolution. This group aimed to strengthen scientific and cultural ties between Weimar Germany and Soviet Russia – both of which were somewhat isolated in the post-World War I geopolitical climate.

Einstein's active participation in this association during a period marked by significant political tension and skepticism toward the Soviet regime underlined his belief in the transformative power of science and culture. By promoting Russo-German exchanges, Einstein sought to facilitate mutual understanding and respect for the emerging scientific and cultural advancements in Soviet Russia, which he considered vital for the progress of civilization. His efforts reflected a steadfast commitment to transcending political and ideological barriers through dialogue and cooperation.

Amid these professional successes, 1923 also saw Einstein's personal life grow increasingly complex when he began a secretive relationship with Betty Neumann, the niece of his close friend Hans Mühsam. This development added a layer of personal intricacy that contrasted sharply with his public image as a revered scientific figure, highlighting the human vulnerabilities even the most esteemed intellectuals face.

Despite the complications arising from Einstein's affair with Neumann, Elsa Löwenthal, his wife, demonstrated remarkable resilience and loyalty. She continued to manage their household and played a critical role in maintaining Einstein's public image, adeptly handling their social affairs and media interactions as his international fame expanded. Elsa's steadfast support was crucial, helping Einstein navigate the complexities of his life during this tumultuous period.

The mid-1920s further showcased Einstein's significant impact on science and the broader socio-political landscape. The year 1924 marked a significant moment in Einstein's scientific career when he collaborated with Indian physicist Satyendra Nath Bose. Together, they developed the concept of Bose-Einstein Condensation. This phenomenon predicted that at extremely low temperatures, particles known as bosons can occupy the lowest quantum state, leading to macroscopic quantum phenomena. This discovery not only expanded the scope of quantum mechanics but also highlighted Einstein's receptiveness to revolutionary ideas, regardless of their geographical origin.

Einstein's engagement with Bose exemplified his broader commitment to international scientific collaboration and dialogue. It was an extension of his spirit that science should serve as a bridge between diverse cultures and nations, fostering mutual understanding and cooperation. This principle was evident in his scientific endeavors and his active participation in global discussions on peace and intellectual exchange.

In 1925, Einstein's role as a global advocate for peace and cooperation deepened significantly. His travels through South America – visiting Argentina, Brazil, and Uruguay – were not just academic missions but also diplomatic endeavors. During these visits, Einstein engaged with local scientists, educators, and political leaders, discussing his theories and advocating for stronger scientific and cultural ties between these nations and Europe. His lectures and public talks emphasized the importance of scientific education and international cooperation in promoting peace and development.

That same year, Einstein's dedication to pacifism deepened significantly when he joined forces with renowned global figures, including Mahatma Gandhi, to sign a manifesto opposing compulsory military service. Mahatma Gandhi, a pivotal leader from India, was widely respected for his philosophy of nonviolent resistance, which he employed to lead the struggle for India's independence from British rule. Gandhi's methods and his steadfast commitment to nonviolence had a profound impact worldwide, resonating with Einstein's own beliefs in peaceful protest and the power of civil disobedience. Together with Gandhi and others, Einstein's action of signing the manifesto symbolized a powerful stand for peace and a collective opposition to the militarism that was spreading through many parts of the world at the time.

This action was aligned with his belief that militarism undermined human welfare and peace. By lending his voice to such causes, Einstein used his global stature to advocate for non-violence and civil responsibility, principles that resonated with his scientific pursuits of understanding the natural world without resorting to destruction.

Einstein's contributions were globally recognized with numerous awards, including the gold medal of the Royal Astronomical Society in London, in 1926. Yet, during this same period, the development of quantum mechanics by Werner Heisenberg, Max Born, and Erwin Schrödinger was causing him significant intellectual unease. Quantum mechanics introduced a framework where probabilities played a central role in predicting physical phenomena. According to this theory, particles do not have definite positions, velocities, or paths until they are observed; instead, they exist in a state of probability until measurement collapses this state into a determinate outcome. This marked a radical departure from classical physics, which posited that objects have definite properties and behaviors that can be calculated with precision given sufficient information.

Einstein's discomfort with quantum mechanics was famously encapsulated in his objection that "God does not play dice with the universe," signifying his resistance to the idea that fundamental aspects of reality are governed by chance. For Einstein, the universe was an ordered and rational place, where events occurred in a predictable pattern that could ultimately be understood through the laws of physics. He believed that if a theory only offered probabilistic predictions, then it indicated a lack of understanding of the underlying realities of nature.

The fifth Solvay Congress in 1927 provided a forum for Einstein to challenge the new quantum theory directly in discussions with its main architects, including Niels Bohr, who was one of its staunchest defenders. The debates between Einstein and Bohr were not merely academic disagreements but were profound engagements that delved into the philosophical underpinnings of reality and knowledge.

During his debates on the foundational concepts of quantum mechanics, Einstein introduced thought experiments designed to illustrate what he saw as significant flaws or paradoxes within the theory. One of the most notable of these was the Einstein-Podolsky-Rosen (EPR) paradox, named after Einstein and his collaborators Boris Podolsky and Nathan Rosen. This paradox challenged the principle of locality and the concept of causality in quantum theory.

In quantum mechanics, locality refers to the idea that objects separated in space cannot instantaneously affect each other's states – a principle rooted in the speed of light limit established by Einstein's theory of relativity. Causality, on the other hand, suggests that an effect cannot occur before its cause. The EPR paradox posits a scenario where two particles interact and then move apart. According to quantum mechanics, even when separated by vast distances, measuring the state of one particle instantly determines the state of the other, suggesting a sort of "spooky action at a distance," as Einstein famously criticized.

This scenario appears to violate locality because the state of one particle affects the state of another instantaneously, regardless of the distance separating them, challenging the relativity-imposed speed limit on how quickly information can travel. It also muddles causality, as it implies an instantaneous relationship between cause and effect across space. Einstein used this paradox to argue that quantum mechanics might be incomplete – potentially missing variables that could provide a more intuitive and complete description of reality.

However, Niels Bohr's responses to Einstein during these debates were crucial in shaping the philosophical acceptance of quantum mechanics. Bohr argued that quantum mechanics was complete and that its probabilistic nature did not reflect a flaw but a fundamental aspect of how physical reality operates at the microscopic level. He introduced the principle of complementarity, suggesting that particles could exhibit characteristics of both waves and particles, depending on the experimental setup, and that these traits were complementary.

The debates had profound implications for the philosophy of science, influencing how generations of physicists and philosophers would come to understand the role of the observer in shaping the outcomes of measurements, the limits of what we can know about the quantum world, and how that knowledge influences the measured phenomena. Despite his reservations, Einstein's challenges to quantum mechanics spurred a deeper examination of the theory, leading to its robustness and the development of more refined theories that continue to hold up under rigorous experimental testing.

By 1928, the cumulative effects of Albert Einstein's extensive commitments – ranging from his deep dives into the complexities of theoretical physics, to his active participation in global intellectual and political dialogues, and his extensive travel schedule – began to manifest in serious health issues. Einstein suffered from a severe heart ailment, a clear signal that the relentless pace of his professional life was unsustainable. This health crisis necessitated a significant slowdown, compelling him to substantially reduce his professional and travel activities to focus on recovery.

During this challenging period, Einstein's personal life underwent a significant change with the arrival of Helen Dukas. Joining his household in 1928, Dukas began her long-standing role as Einstein's secretary, and she eventually became an indispensable figure in his life, later taking on the additional duties of a housekeeper. Born in 1896 in Germany, Helen Dukas was initially hired to manage Einstein's increasing correspondence and the increasingly complex demands of his public and professional life, but her role quickly expanded due to Einstein's health issues and his need for a more comprehensive support system.

Helen Dukas proved to be much more than a secretary; she was a fiercely loyal protector of Einstein's privacy, a meticulous manager of his professional affairs, and a close confidante. Her responsibilities included managing his appointments, handling his correspondence with scientists, students, and public figures, and ensuring that his manuscripts and publications were organized and accessible. She also played a critical role in managing the vast amount of fan mail and numerous requests that Einstein received regularly.

Moreover, Dukas was instrumental in helping Einstein navigate the complexities of life as a public figure. She screened his calls and visitors, guarded him against unwelcome intrusions, and helped maintain a stable and quiet environment in which he could work. Her commitment went beyond mere professional duty; she was deeply dedicated to Einstein's well-being, advocating for his health and privacy.

Despite health challenges, the late 1920s were marked by continued honors and an expanding role as a public intellectual. Einstein's 50th birthday in 1929 was celebrated with the construction of a summer house in Caputh near Potsdam, symbolizing his deep roots in Germany, even as the political climate there grew increasingly hostile.

As the decade closed, Einstein not only revisited his pacifist stance by signing a manifesto demanding global disarmament but also prepared for another significant chapter in his life with a second visit to the U.S. His journey to the California Institute of Technology marked the beginning of what would become an extended stay in the United States, driven both by his scientific pursuits and the deteriorating political situation in Europe.

As the 1920s drew to a close, Albert Einstein found himself navigating significant personal and professional milestones that indicated the coming complexities of the 1930s. The decade had ushered in profound achievements and notable celebrations, including his 50th birthday, but it was also a period marked by escalating political tensions and impending global upheavals. These changes foreshadowed a dramatic shift in Einstein's life and his eventual disconnection from his homeland, setting the stage for his future in America.

During this time, Einstein's engagement with the Solvay Congress in the late 1920s proved to be a momentous aspect of his career – not only for his contributions to the development of quantum physics and his famous debates with other leading scientists but also for fostering relationships with influential figures outside the scientific sphere.

Among these was his notable friendship with Queen Elisabeth of Belgium, a relationship that blossomed around their shared intellectual passions and commitment to humanitarian and cultural advancement. Queen Elisabeth, a member of the Bavarian royal family and married to King Albert I of Belgium, was an avid supporter of the arts and sciences and brought a profound depth of curiosity and engagement that resonated deeply with Einstein.

Their connection, established during the intellectually charged atmosphere of the Solvay Congress, exemplified the confluence of science and broader cultural dialogue. Einstein found in Queen Elisabeth not only an admirer of his scientific work but also a keen conversationalist who was eager to delve into the philosophical and cultural implications of new scientific discoveries. Their discussions transcended the boundaries of physics, touching on philosophy, music, and the pressing political issues of the day, reflecting Einstein's broader view of science as a bridge linking diverse realms of human thought and experience.

This period also saw continued recognition of Einstein's scientific contributions, as evidenced by the honors he received. Albert Einstein received the Max Planck medal in 1929. This award, given by his mentor and friend Max Planck, along with an honorary doctorate from the University of Paris, were not just acknowledgments of his past contributions to physics but also affirmations of his ongoing impact on the global scientific community. These honors underscored his stature as one of the leading minds of his time, capable of drawing both scientific and cultural circles together through his groundbreaking theories and his compelling vision for a more interconnected world.

Together, these elements from the late 1920s painted a picture of a man at the zenith of his scientific influence, actively engaging with the intellectual and cultural leaders of his time, and poised on the brink of a new chapter that would challenge his convictions and redirect his life's work in response to the shifting tides of history.

The year 1930 was a pivotal one for Albert Einstein, marking significant developments in both his personal life and his broader philosophical and professional engagements. This year was woven with threads of familial joy, intensified political activism, and deepening academic commitments, reflecting a holistic integration of Einstein's values and experiences that influenced his trajectory both as a scientist and a global public figure.

Hans Albert Einstein, who became a distinguished engineer and professor of hydraulic engineering, maintained a close and affectionate relationship with his father, Albert Einstein, despite being geographically distant. They communicated regularly through letters, discussing family matters, personal concerns, and occasionally physics. The birth of Hans Albert's son, Bernhard Caesar Einstein, brought immense joy to Albert Einstein, marking a significant moment of personal happiness and continuity in his life. The arrival of his grandson not only deepened Einstein's happiness but also provided a sense of legacy amidst the growing troubles in Europe. As a grandfather, Einstein embraced a new identity that enriched his familial ties and complemented his ongoing contributions to science, public affairs, and his dedication to peace.

1930 also saw Einstein intensifying his commitment to global peace and disarmament, a cause he had supported strongly since the aftermath of World War I. His pacifism was put into action when he signed a manifesto demanding the world's disarmament, an appeal made against the backdrop of rising militarism and nationalism in various parts of the world, particularly in his native Germany. This act was symbolic of Einstein's growing engagement with political causes, standing as a public intellectual who was not only concerned with theoretical physics but also deeply invested in the ethical implications and societal impacts of science and technology.

Einstein's professional life during this year also took a significant turn as he returned to the United States for a research stay at the California Institute of Technology (Caltech) in Pasadena. This visit marked his second to the United States, a country that would soon become his permanent residence in light of the escalating political situation in Germany. His time at Caltech was not just a routine academic visit but a meaningful engagement with some of the leading scientific minds in America.

During his visit to the United States in the early 1930s, Albert Einstein was met with enthusiastic attention, reflective of his stature as a global scientific icon. Caltech facilitated his stay, mindful of his previous overwhelming experiences with the media in 1921, thus declining numerous invitations on his behalf that required public appearances. However, Einstein graciously allowed some interactions with his admirers, balancing his need for privacy with his fans' desire to engage with him.

Upon his arrival in New York City, Einstein's schedule was packed with various culturally enriching activities. Einstein was escorted on a tour of Chinatown, immersing himself in the cultural diversity of the area. He also shared a meal with the editors of The New York Times and attended a performance of "Carmen" at the Metropolitan Opera, where he was warmly received by the audience, showcasing his popularity and the public's enthusiasm for his presence. The city welcomed him wholeheartedly, with Mayor Jimmy Walker presenting him with the keys to the city, symbolizing his honorary embrace by New York. His interactions included meeting notable figures such as Nicholas Murray Butler, the president of Columbia University, who praised Einstein as "the ruling monarch of the mind," and Harry Emerson Fosdick, who showed him a statue erected in his honor at Riverside Church.

Einstein's journey continued to California, where his visit to Caltech underscored differing viewpoints with Caltech president Robert A. Millikan, particularly on issues of militarism, reflecting Einstein's staunch pacifist stance. He openly critiqued the misuse of science for destructive ends during a talk to Caltech students. During his stay, he also developed relationships with cultural figures such as Upton Sinclair and Charlie Chaplin, both known for their pacifist views. Chaplin, in particular, formed a close friendship with Einstein, inviting him and his wife, Elsa, to his home and later to the Hollywood premiere of "City Lights," one of the era's most anticipated events.

Einstein's musical activities also continued during his academic tenure in the United States. While at the California Institute of Technology in 1931, he visited the Zoellner family conservatory in Los Angeles, where he performed pieces by Beethoven and Mozart with members of the Zoellner Quartet*. Music remained a source of comfort and expression for Einstein until the end of his life. Notably, when the young Juilliard Quartet visited him in Princeton, he played his violin with them, impressing them with his coordination and intonation.

These experiences in the United States emphasized Einstein's broad appeal, both as a scientist and as a public figure whose influence transcended academic circles. His interactions during this trip not only reinforced his connections with the American scientific and cultural communities but also provided a backdrop to his evolving views on the cosmological constant.

Initially introduced by Einstein in 1917 to maintain a model of a static universe, the cosmological constant (denoted as Lambda, λ) faced new challenges following Edwin Hubble's revelation of an expanding universe. The observations of galaxies moving away from each other contradicted the static universe model and suggested a dynamic, expanding cosmos. This shift was supported by other cosmologists, including Georges Lemaître, who proposed models that did not necessitate a cosmological constant, prompting Einstein to reconsider and eventually discard this aspect of his theory as the "biggest blunder" of his career. His experiences and discussions during his travels undoubtedly played a role in his continual reevaluation of his theoretical frameworks, reflecting his commitment to adapting his theories in light of new scientific evidence.

During his visit to Caltech and other American institutions, Einstein engaged deeply with other leading physicists and astronomers,

---

* The Zoellner Quartet was a renowned string quartet active in the early 20th century. It was founded by Adolf Zoellner, a violinist, and comprised members of his family. The quartet was well-known in the United States and played a significant role in popularizing chamber music there. They were celebrated for their performances and contributed to the musical culture of the era, often collaborating with prominent musicians and composers.

discussing the latest developments and implications of the expanding universe theory. These discussions prompted Einstein to reconsider the necessity and validity of the cosmological constant he had once staunchly defended.

Einstein's decision to remove the cosmological constant from his equations was not merely a technical adjustment but a profound shift in his understanding of the universe. This moment of self-correction reflected his intellectual honesty and his willingness to admit errors – an approach that is crucial for scientific progress but often difficult to embrace.

Einstein's willingness to discard the cosmological constant had significant implications for the field of cosmology, freeing researchers to explore new models of the universe that aligned more closely with observational data. It underscored the importance of empirical evidence in shaping theoretical constructs in physics and highlighted the dynamic nature of scientific inquiry, where theories are not static but evolve in response to new data. Furthermore, Einstein's change of stance exemplified his broader scientific philosophy that theories must be adaptable and responsive to the physical realities they aim to describe.

By 1932, Einstein was deeply entangled in the academic life of the United States, having accepted a position at the newly founded Institute for Advanced Study in Princeton, New Jersey. This arrangement was initially intended to be part-time, with Einstein planning to split his year between Princeton and Berlin. However, the rise of the Nazi regime in Germany dramatically altered these plans. The political and personal threats posed by the Nazis, especially given Einstein's Jewish heritage and his outspoken political views, made it impossible for him to continue living in Germany.

In early 1933, Albert Einstein was concluding a visiting professorship at the California Institute of Technology in Pasadena when he became acutely aware of the dire situation unfolding back in Germany. During his absence, the Gestapo had conducted multiple raids on his family's apartment in Berlin, an ominous sign of the escalating hostility toward individuals deemed undesirable by the new regime. The Gestapo, short for Geheime Staatspolizei (Secret State Police), was the official secret police of Nazi Germany. Formed in 1933, it was tasked with suppressing opposition to the Nazis through terror, intimidation, and arrest, often targeting Jews, political dissidents, and others considered threats to the state ideology.

By March, as Einstein and his wife Elsa prepared to return to Europe, they were confronted with alarming news: the German Reichstag (parliament) had passed the Enabling Act* on March 23, granting Adolf Hitler the powers of a legal dictatorship. This critical development underscored the perilous state of affairs in Germany, particularly for prominent Jews like Einstein.

The situation deteriorated further when Einstein learned that not only had his Berlin apartment been targeted, but their cottage had also been raided and his personal sailboat confiscated by the Nazis. These personal violations were compounded by the transformation of his cottage into a Hitler Youth camp, a symbolic repurposing that starkly illustrated the regime's intent to erase Jewish presence and influence.

Upon their arrival in Antwerp, Belgium, on March 28, Einstein took a decisive step by visiting the German consulate to surrender his passport, formally renouncing his German citizenship. This act marked a definitive divorce from his homeland in response to the oppressive and dangerous political environment fostered by the Nazis.

This period marked a profound turning point in Einstein's life.

---

* The Enabling Act, officially known as the "Law to Remedy the Distress of the People and the Reich," was passed on March 23, 1933. This legislation effectively gave Adolf Hitler and his government the authority to enact laws without the involvement of the Reichstag, Germany's parliament. It marked a significant step in the Nazi rise to power, as it allowed Hitler to establish a legal dictatorship by bypassing the legislative checks and balances, consolidating his control, and enabling the implementation of policies without parliamentary consent or constitutional oversight. This act was one of the key events that dismantled the democratic structures of the Weimar Republic, leading to the totalitarian regime of Nazi Germany.

Recognizing the severe implications of Hitler's rise to power, particularly the aggressive anti-Semitic policies that now characterized the government, Einstein decisively broke all ties with Germany. His renunciation of German citizenship was not merely a personal safeguard but also a strong political statement against the regime that threatened the very principles he stood for. Furthermore, his resignation from the Prussian Academy of Sciences and his renunciation of the Order "Pour le Mérite" were symbolic gestures that reflected his broader disapproval of the Nazi infiltration of Germany's cultural and scientific institutions. These actions underscored his rejection of the extremely authoritarian ideology and his commitment to standing against the corruption and moral decay it represented, setting the stage for his future advocacy and life in exile.

In April 1933, Albert Einstein was confronted with the harsh realities of the new German government's policies when laws were enacted barring Jews from holding official positions, including academic posts. Historian Gerald Holton notes that this was met with virtually no protest from non-Jewish colleagues, leading to the abrupt dismissal of thousands of Jewish scientists from their university roles. Their names were taken out from institutional rosters in a sweeping act of exclusion.

A month after these academic purges, the German Student Union escalated the attack on Jewish intellectualism by targeting books written by Jews and other "undesirables" in large public burnings. During these events, Joseph Goebbels, the Nazi propaganda minister, infamously declared, "Jewish intellectualism is dead." The targeting of Einstein's works during these burnings underscored the regime's hostility toward him. This animosity was further accentuated by a German magazine that ominously listed him among the enemies of the state, alarmingly noting that he was "not yet hanged" and placing a bounty on his head.

Einstein, deeply shaken by these events, expressed his shock and dismay in a letter to his friend, physicist Max Born, who had already fled to England. He remarked on the unexpected level of brutality and cowardice shown by the Nazis. After relocating to the United

States, he referred to the book burnings as a "spontaneous emotional outburst" by those fearing the influence of intellectually independent individuals.

Now stateless and without a permanent home, Einstein's concerns extended beyond his own safety; he was also worried about the fate of many other scientists still trapped in Germany. With the assistance of the Academic Assistance Council, an organization founded to help academics escape Nazi persecution, Einstein found temporary refuge in Belgium. Later, he visited England for about six weeks at the invitation of Commander Oliver Locker-Lampson, a British MP* who became a close friend and supporter. During his stay, Locker-Lampson arranged for Einstein to meet influential figures such as Winston Churchill, Austen Chamberlain, and former Prime Minister Lloyd George to advocate for the rescue of Jewish scientists from Germany.

Churchill responded promptly, instructing his physicist friend Frederick Lindemann to help place Jewish scientists in British universities. This initiative significantly aided the Allies' technological advancements during World War II. Moreover, Einstein's advocacy efforts extended beyond Britain; he wrote to other world leaders, including Turkey's Prime Minister İsmet İnönü, requesting assistance for unemployed German-Jewish scientists. His efforts led to over a thousand Jewish intellectuals finding refuge in Turkey.

During his time in England, Locker-Lampson also proposed a bill to grant British citizenship to Einstein. Although this bill failed, it highlighted Einstein's status as a global citizen and his advocacy for Jewish rights during a period of increasing peril. Despite the failure of the citizenship bill, Einstein continued to speak out against the injustices occurring in Germany, and he was eventually offered a position at the Institute for Advanced Study in Princeton, New Jersey, where he would spend the remainder of his career. He moved to the United States in 1933.

In Princeton, Einstein found an environment compatible with his work and thoughts. The Institute for Advanced Study, located in

---

* MP stands for member of Parliament.

Princeton, New Jersey, was established in 1930 as an academic institution dedicated to the pursuit of advanced learning and research across a spectrum of disciplines. This institution was designed to offer an environment where intellectual exploration could occur unimpeded by the typical demands of teaching obligations found at traditional universities. This unique academic setting provided scholars with the freedom to focus fully on research, encouraging deep dives into complex and innovative areas of study.

At the Institute, Einstein was able to continue his exploration of theoretical physics without the distraction of classroom duties. This allowed him significant time to focus on developing his unified field theory – an ambitious goal that aimed to describe all fundamental forces of nature. Moreover, Princeton offered Einstein a platform from which he could voice his concerns about the political developments in Europe and advocate for peace and rational governance. Despite his physical removal from Europe, Einstein remained deeply engaged with the issues affecting the continent, using his status to influence public opinion and policy, particularly in advocating for refugee assistance and opposing the growing militarism of the Nazi regime.

The mid to late 1930s represented a period of profound personal challenges and significant professional achievements for Albert Einstein. This era was punctuated by notable losses that deeply affected him, but also by enduring contributions to the field of physics, reflecting his resilience and unwavering commitment to his work.

1936 was particularly difficult for Einstein. He suffered the loss of his second wife, Elsa, who had been a constant companion and support since their marriage in 1919. Elsa had been ill for some time, and her death left a significant void in Einstein's personal life. Her role had been pivotal, not only in managing his domestic affairs but also in providing emotional support through the tumultuous years of their life together, including their emigration to the United States. Elsa was with Einstein in the U.S. when she died, at the age of sixty. She consistently accompanied him during their escape from the rising dangers of World War II in Europe. Elsa's passing marked the end of a profound partnership, forcing Einstein to adjust to a new personal landscape.

That same year, Einstein also mourned the death of Marcel Grossmann, a close friend and former classmate from his days at the Polytechnic Institute in Zurich, who passed away in Zurich, Switzerland. Grossmann was instrumental in the development of the general theory of relativity; it was he who introduced Einstein to the tensor calculus, which played a major role in the formulation of the theory. Grossmann's death was not just a personal loss for Einstein but also a reminder of the fleeting nature of the collaborations that had shaped his earlier scientific career.

Despite these personal sorrows, Einstein continued to advance his scientific work with undiminished vigor. In 1938, along with his colleague Leopold Infeld, Einstein published "The Evolution of Physics." This work was significant not only for its content but also for its accessible approach to explaining complex scientific concepts. The book traced the development of physics from the early concepts of classical mechanics introduced by Isaac Newton to the more recent advances in quantum theory. It aimed to provide a clear understanding of how the field of physics had evolved and how the fundamentals of science were interconnected across different theories.

"The Evolution of Physics" served several purposes. It was an educational tool that demystified complex theories for a broader audience, embodying Einstein's belief in the importance of making scientific knowledge accessible and understandable to the public. Moreover, the book emphasized Einstein's view of scientific theories as part of an evolving continuum rather than as isolated discoveries. This perspective was crucial during a time when the foundations of physics were being dramatically reshaped by new theories and discoveries.

In 1939, as the shadow of World War II loomed over the world, Albert Einstein took one of the most significant actions of his life, transcending his usual realms of theoretical physics and academic discourse. Understanding the grave implications of nuclear physics in the context of war, Einstein chose to step directly into the political arena. He signed a letter to President Franklin D. Roosevelt that warned him of the potential development of "extremely powerful bombs of a new type," an action that played a major role in the U.S. government's decision to invest in nuclear research and ultimately led to the establishment of the Manhattan Project – a top-secret research and development project during World War II that produced the first nuclear weapons.

The letter to President Franklin D. Roosevelt was influenced by conversations with physicist Leo Szilard, among other European refugee scientists. These scientists were alarmed by recent advancements in nuclear physics, particularly nuclear fission, a process discovered in uranium. This process, identified by scientists in Berlin, involves the splitting of an atomic nucleus into smaller parts, releasing a significant amount of energy. This discovery was monumental because it indicated the potential to harness this energy for massive power outputs, which could, unfortunately, include the creation of powerful weapons of mass destruction.

Given the tense geopolitical climate and the potential for Nazi Germany to utilize this technology for war, Szilard and his colleagues felt a pressing need to inform and mobilize U.S. leadership. They chose to approach Albert Einstein, whose scientific prestige and personal convictions against the Nazi regime positioned him as the ideal figure to raise the alarm. Szilard believed that a letter from Einstein, given his renowned status and influence, would be taken seriously by government officials and prompt them to consider immediate action to counteract the potential threat. This led to the drafting of a letter that Einstein would send to President Roosevelt, highlighting the dangers and implications of nuclear weapons development and urging the United States to accelerate its own research into nuclear fission.

Einstein was already well known for his pacifist views, but the urgent threat posed by the possibility of Nazi Germany developing a nuclear weapon convinced him of the need for preemptive action. His letter to Roosevelt outlined the new advancements in nuclear chain reactions and urged the necessity of U.S. attention and involvement in nuclear research. Einstein's letter explained that a "single bomb of this type, carried by boat and exploded in a port, might very well destroy the whole port together with some of the surrounding territory." Such a stark warning underscored the catastrophic potential of atomic energy if used for warfare.

The letter was impactful, leading Roosevelt to initiate the Advisory Committee on Uranium in October 1939, which eventually evolved into the Manhattan Project. Einstein himself did not participate in the development of nuclear weapons, and he was kept unaware of the specifics of the Manhattan Project. However, his foundational work in physics had indirectly made such developments possible.

Einstein later reflected on the decision to send the letter with mixed feelings. After the war, he expressed regret about the role his letter played, particularly following the atomic bombings of Hiroshima and Nagasaki. He became a strong advocate for nuclear disarmament and greater international cooperation to prevent nuclear war.

As World War II began with the German invasion of Poland in September 1939, Einstein's advocacy for peace and disarmament took on new urgency. His experiences throughout the 1930s, from the rise of fascism in Europe to his resettlement in America, underscored a decade of profound change, during which Einstein's legacy as both a scientist and a humanitarian was firmly cemented.

Years of Reflection

Albert Einstein's final decade in America was not only a period of profound personal transitions but also a time when his scientific legacy profoundly intersected with global events. From becoming a U.S. citizen to witnessing the devastating use of atomic energy during World War II, Einstein's experiences from 1939 to 1949 reflected his complex relationship with his own discoveries and his deepening commitment to peace and humanitarianism.

In 1940, amidst a backdrop of escalating global conflict, Albert Einstein took a deeply symbolic and transformative step in his personal life by swearing the oath on the American Constitution, officially becoming a U.S. citizen. This significant event was not merely a bureaucratic formality but a profound declaration of allegiance and a commitment to the United States, a nation that had provided refuge from the growing threat of Nazism in his native

Germany. Retaining his Swiss citizenship, Einstein embraced a dual identity that reflected his European roots and his new American ties, encapsulating his complex relationship with national identity.

Einstein's decision to take American citizenship was influenced by several factors. Since his relocation to the United States in 1933, following Adolf Hitler's rise to power in Germany, Einstein had found a vibrant intellectual community at Princeton, New Jersey, where he was warmly welcomed and given the freedom to continue his research unimpeded by the oppressive policies of the Nazi regime. The academic environment and the broader political spirit of freedom and democracy in the U.S. resonated with Einstein's own ideals, which were deeply rooted in his advocacy for civil liberties and justice.

Moreover, becoming an American citizen during World War II carried significant weight. The world was involved in a war that was, in many respects, a fight against the fascist ideologies that Einstein vehemently opposed. By adopting American citizenship, Einstein symbolically aligned himself with the forces opposing Nazi Germany, marking his support for the Allies' efforts in what he saw as a moral and existential battle against tyranny and oppression.

This act of citizenship also underscored Einstein's gratitude toward the United States for its role as a sanctuary for himself and many other European intellectuals and scientists fleeing persecution. His decision was both a personal milestone and a public statement, reinforcing his belief in American principles and his hope that the U.S. would continue to be a leader in the fight for a more just and peaceful world.

Additionally, by maintaining his Swiss citizenship, Einstein preserved a vital connection to Europe and his past. This link was not just about identity but also a practical acknowledgment of the complex interplay of culture, history, and personal allegiance that defined his life. The dual citizenship thus represented a bridge between his European origins and his American present, embodying his broader view of a connected and interdependent world.

The year 1939 had already set a dramatic context for his naturalization: it was when Einstein signed his famous letter to President Franklin D. Roosevelt, warning of the potential for nuclear weapons development. This letter would catalyze the U.S. government's investment in atomic research.

Despite Einstein's crucial role in prompting U.S. atomic research, he was excluded from direct involvement in the Manhattan Project. There were several reasons for this decision. Firstly, Einstein was known for his pacifist leanings and political activism, which could have raised concerns among military and governmental leaders managing a highly classified, strategic war project. Moreover, his status as a prominent refugee scientist from Germany, coupled with his associations with various international peace movements, may have contributed to a perception of him as a security risk.

However, Einstein's exclusion from the Manhattan Project did not entirely sideline him from contributing to the war effort. In 1943, recognizing his vast knowledge and expertise in related fields, the U.S. Navy employed Einstein as an advisor for highly explosive materials. This role, although distinct from the core atomic weapons development, leveraged his deep understanding of physics and its practical applications in other areas of military technology.

In this capacity, Einstein was able to contribute to the war effort, focusing on problems related to the behavior and stability of various explosive compounds under different conditions. His work helped in enhancing the safety and effectiveness of explosive materials that were critical to the war, demonstrating that his expertise could still be utilized constructively without directly involving him in the more controversial aspects of atomic weapons development.

This arrangement emphasized the nuanced ways in which Einstein's capabilities were recognized and utilized during the war. While he was kept at a distance from the most secret elements of the Manhattan Project, his advisory role with the Navy signified a compromise – balancing the need for security with the desire to harness his brilliance for the Allied cause.

The dropping of atomic bombs on Hiroshima and Nagasaki in August 1945 marked a profound turning point for Albert Einstein. The unprecedented destruction wrought by these weapons deeply troubled him, as the theoretical work he had pioneered in the early 20th century had now been used to unleash devastating power. This was a realization of his theories' potential for harm at a scale and magnitude he had never intended, prompting a significant reevaluation of his stance on nuclear technology and its control. His response to these events reshaped his post-war agenda, focusing heavily on advocating for nuclear disarmament and the peaceful use of atomic energy.

Following the conclusion of World War II, Albert Einstein, deeply shaken by the destructive capabilities of atomic weapons, increasingly viewed his role through a broader, more globally responsible lens. Recognizing the potentially catastrophic consequences of nuclear technology, Einstein devoted much of his post-war efforts to advocating for substantial political changes on an international scale.

In 1946, Albert Einstein took decisive action by writing an open letter to the newly established United Nations, advocating for the creation of a world government. He argued passionately that the concept of national sovereignty, which allows countries to govern themselves independently, could potentially lead humanity toward catastrophic conflicts, especially with the emergence of nuclear weapons. According to Einstein, the competitive drive and power struggles inherent in national governments made them prone to engage in conflicts that could escalate into global destruction.

Einstein proposed that the only viable solution to prevent such disasters was the establishment of a supranational government – a governing body above individual nations – equipped with legislative powers to make laws and enforcement powers to implement them. This global authority would be responsible for maintaining worldwide peace and security by regulating and controlling the use of nuclear energy and other global issues that transcend national borders.

He envisioned this world government not as a power unto itself, but as a framework for international cooperation and conflict resolution, where the collective interests of humanity take precedence over individual national interests. Einstein believed that the United Nations could serve as the foundation for this broader, more powerful entity, but he was also critical of the UN's structure at the time. He pointed out that its effectiveness was limited by the veto power held by the major nations, which could block substantial actions necessary for global peace and security.

This belief in a world government was not a passing interest for Einstein; rather, it was a deep-seated conviction. He viewed it as crucial for humanity's survival and prosperity in an era where technological advancements, especially in weaponry, could potentially lead to unparalleled destruction. Einstein's advocacy for this idea stemmed from his broader view of politics and his commitment to pacifism, reflecting his profound concern for the future of mankind in the nuclear era.

Einstein's commitment to ensuring the responsible use of atomic energy led him to take on an influential role as the head of the Emergency Committee of Atomic Scientists in 1946. This committee was composed of leading physicists who had witnessed first-hand the development and deployment of nuclear weapons. Under Einstein's leadership, the committee worked tirelessly to educate the public and political leaders about the dangers of atomic weapons and the potential benefits of nuclear energy if used for peaceful purposes.

The committee's efforts included public speeches, open letters, and private lobbying, aiming to promote legislative and international measures to control and regulate atomic energy. Einstein's fame and credibility brought significant attention to these issues, helping to elevate them in the public and political discourse.

During the late 1940s, alongside his deepening engagement with global political issues and his advocacy for peace, Albert Einstein also encountered several profound personal challenges that marked this period as one of intense personal hardship and reflection.

In 1948, Einstein faced a significant health crisis when he was diagnosed with an aortic aneurysm, a serious condition involving a dilation of the aorta, the main artery carrying blood from the heart to the rest of the body. This diagnosis was particularly alarming because if the aneurysm ruptured, it could lead to life-threatening complications. The condition necessitated immediate surgical intervention. Einstein underwent an innovative surgical procedure involving the application of a synthetic film (cellophane) to wrap the aorta and reinforce it. This surgery, experimental at the time, was a risk but ultimately successful, allowing Einstein to continue his work and advocacy with the caution of needing regular medical monitoring and a significantly modified lifestyle to manage his condition.

The same year also brought personal losses that deeply affected Einstein. His first wife, Mileva Maric, with whom he had shared a complex and often strained relationship since their separation in 1919, passed away in August 1948. Although their marriage had ended many years before, the news of her passing undoubtedly evoked a complex mix of emotions, including reflections on their early years together when they were both passionate about physics and their difficult later years.

Moreover, Einstein's sister, Maja, whom he was very close to throughout his life, suffered a debilitating stroke. Maja had lived with Einstein in Princeton after fleeing Europe, and her illness represented not only a personal sorrow for Einstein but also a poignant reminder of the vulnerabilities associated with aging. Her condition required extensive care and significantly impacted Einstein's emotional well-being, as he deeply cared for his sister.

These personal health and family challenges occurred against the backdrop of Einstein's vigorous public engagement with issues of global significance, such as nuclear disarmament and the promotion of peace. Balancing these public responsibilities with his personal struggles required considerable resilience and highlighted the complex interplay between his private life and his public persona. Despite these challenges, Einstein continued to work and advocate for his beliefs with undiminished passion.

In 1949, amidst a period marked by personal recovery and reflection, Albert Einstein published *Autobiographical Notes*, a profound introspective work that allowed him to convey his thoughts on a lifetime of scientific inquiry and the philosophical lessons derived from his experiences. This publication not only chronicled his scientific journey but also delved deeply into the philosophical and ethical dimensions of his work, providing a unique window into the mind of one of the 20th century's greatest thinkers.

Thus, the decade from 1939 to 1949 was one of immense change and reflection for Einstein. As he adapted to life in America, he witnessed the global impact of his scientific theories and grappled with the ethical responsibilities they entailed. His advocacy for peace, his reaction to the uses of atomic energy, and his personal losses and health challenges defined this period. Through it all, Einstein remained deeply engaged in the world around him, committed to the idea that science should serve humanity's broader goals of peace and progress.

The Final Years

As the 1950s dawned, Albert Einstein approached the final decade of his life – a period marked not only by continued intellectual activity but also by significant personal reflections and decisions that would shape his enduring legacy. The decade was punctuated by poignant moments that encapsulated his philosophical, scientific, and humanitarian concerns.

In March 1950, Albert Einstein took a significant step to secure his intellectual and personal legacies by signing his last will. This document was more than a mere legal formalization; it was a deliberate and thoughtful arrangement that reflected his values and priorities. By appointing Dr. Otto Nathan, an economist and close friend, and Helen Dukas, his trusted secretary who had managed his professional and personal affairs for many years, as his executors, Einstein ensured that his legacy would be managed by those who understood not only his work but also his ethical and philosophical stances.

Einstein's decision to bequeath his written works and intellectual heritage to the Hebrew University of Jerusalem was particularly indicative of his deep-rooted connections to education and the Jewish community. Founded in 1918 and opened in 1925 with Einstein's active participation, the Hebrew University was a project close to his heart, embodying his hopes for the cultural and intellectual resurgence of the Jewish people. This endowment was not merely an act of donating academic materials; it was a profound endorsement of the university's mission and a reflection of his commitment to the advancement of knowledge. Einstein saw education as a powerful tool for promoting peace and mutual understanding, and he viewed the university as a beacon of these ideals in the tumultuous landscape of the Middle East. His contribution aimed to ensure that future generations could access his scientific and philosophical insights, potentially inspiring new ideas and initiatives that would align with his vision of a harmonious world.

Furthermore, Einstein's legacy planning reflects his broader philosophical views about the responsibility of scientists and intellectuals to contribute positively to society. His life was marked by a consistent effort to align his scientific pursuits with his moral values, advocating for peace, civil rights, and international cooperation. By securing his papers with an institution dedicated to higher learning and intercultural dialogue, Einstein was reinforcing his belief in the power of education to bridge divides and foster a more understanding and cooperative global community.

That same year, Einstein published "Out of My Later Years," a collection that brought together a variety of his non-scientific essays and speeches from the previous two decades. This compilation showcased his broad range of interests and his public engagement on issues beyond theoretical physics, including peace, education, and civil rights. The essays provided insights into his thoughts on humanity, his hopes for global peace, and his concerns about the future in the atomic age.

In June 1951, Einstein faced personal sorrow as his sister, Maja, passed away in Princeton. Maja had been a close confidante throughout his life, and her death marked the end of a deeply personal connection that had provided him with much-needed emotional support, especially during their shared years in exile from Europe.

In the following year came the offer of the presidency of Israel, which was a significant testament to Albert Einstein's international stature and his importance to the Jewish community worldwide. Following the death of Chaim Weizmann, the first President of Israel and a fellow scientist, the newborn state sought a successor who could embody the nation's ideals and aspirations on the global stage. Einstein, renowned not only for his groundbreaking contributions to physics but also for his outspoken support for Zionism and humanitarian issues, was a natural choice. This offer highlighted the profound respect and admiration he commanded.

Einstein's decision to decline the presidency was rooted in several personal and professional reasons. Firstly, he humbly cited his lack of experience in political matters. Known for his integrity, Einstein was keenly aware of the complexities and responsibilities of such a role and felt that his skills as a scientist did not necessarily equip him to manage the intricacies of state governance. He believed that the leadership required to navigate the political landscape, especially in a state as young and as surrounded by geopolitical tensions as Israel, demanded expertise and experience that he did not possess.

Secondly, Einstein's refusal also reflected his commitment to his scientific pursuits. Despite his advanced age and declining health, Einstein remained deeply engaged in his work, particularly in his quest for a unified field theory that he hoped would integrate all the fundamental forces of nature into a single theoretical framework. Accepting the presidency would have required him to abandon these pursuits, something he was unwilling to do. Einstein's identity was deeply tied to his role as a scientist, and his ongoing projects were more than just work; they were a lifelong passion that he was not ready to set aside.

Moreover, Einstein valued his independence as a public intellectual, a role that allowed him to speak and act freely on various social, political, and scientific issues without the constraints typically imposed on a head of state. Throughout his life in the public eye, Einstein had utilized his platform to advocate for issues like pacifism, civil rights, and international cooperation. Becoming a political leader would have placed him in a position where his actions would be dictated by national interests and political considerations, potentially limiting his ability to advocate freely on global issues.

This commitment to independence was particularly evident in 1954, during a period marked by intense Cold War tensions and domestic political paranoia in the United States. At this time, Albert Einstein found himself once again at the center of a significant controversy. This controversy involved his public support for J. Robert Oppenheimer, a pivotal figure in modern physics and the scientific director of the Manhattan Project, which developed the atomic bomb during World War II. After the war, Oppenheimer became a target of the U.S. government's anti-communist witch hunts, led by figures such as Senator Joseph McCarthy, who questioned Oppenheimer's political affiliations and his loyalty to the United States. Einstein's defense of Oppenheimer was consistent with his longstanding commitment to personal and intellectual freedom, reflecting his deep-seated values against political persecution and his advocacy for scientific and moral integrity.

The controversy surrounding J. Robert Oppenheimer, which intensified during the early 1950s, stemmed largely from his previous associations with left-wing movements and his vocal opposition to certain U.S. military policies during the Cold War. As a prominent physicist and the scientific director of the Manhattan Project, Oppenheimer had played a significant role in the development of the atomic bomb. However, after the war, his perspective shifted; he began to express serious ethical concerns about the nuclear arms race and particularly opposed the development of the hydrogen bomb, a weapon far more destructive than the atomic bombs dropped on Japan.

Oppenheimer's political affiliations during the 1930s, when he had connections to various communist and socialist groups, became a focal point for criticism in the McCarthy era – a time characterized by intense fear of communism and widespread paranoia about internal subversion in the United States. These associations, coupled with his later stance on nuclear weapons, made him a target for those who questioned his loyalty and his suitability for holding a position of significant influence in scientific and defense-related matters.

This backdrop set the stage for a fraught confrontation between Oppenheimer and government officials, fueled by the broader context of the Cold War where the U.S. was deeply invested in maintaining and expanding its nuclear arsenal. Oppenheimer's opposition to the hydrogen bomb was seen not just as a policy disagreement but as an ideological betrayal by those who advocated for a strong defense posture against the perceived communist threat. This tension ultimately led to a security clearance hearing that aimed to discredit him and remove him from positions of power within the scientific community, reflecting the era's conflated concerns about security, governance, and ideological purity. In 1954, these issues culminated in a highly publicized security clearance hearing that sought to label Oppenheimer as a security risk, jeopardizing his career and reputation.

Albert Einstein's defense of Oppenheimer was not merely an act of loyalty to a fellow physicist but also a principled stand against what he perceived as the misuse of political power to stifle intellectual freedom and personal integrity. Einstein was acutely sensitive to the dangers of political persecution, having fled Nazi Germany in the 1930s due to similar tactics employed by the Hitler regime against academics and intellectuals. His support for Oppenheimer was thus rooted in a broader critique of the era's political paranoia – often referred to as McCarthyism – under which numerous Americans were unfairly accused of subversion or disloyalty without proper regard for evidence or due process.

Einstein publicly expressed his disdain for the actions taken against Oppenheimer, seeing them as antithetical to the principles of justice and academic freedom that he held dear. He viewed the attacks on Oppenheimer as part of a larger troubling trend toward authoritarianism in American political life, where the freedoms of thought and expression were under threat. By defending Oppenheimer, Einstein sought to uphold the values of open inquiry and dialogue that are essential to scientific progress and democratic society.

In 1954, amidst the personal and professional challenges Albert Einstein faced, his health began to seriously decline, marking a poignant phase in the final chapter of his life. Diagnosed with hemolytic anemia, a condition where red blood cells are destroyed faster than they can be made, Einstein experienced increased frailty that affected his capacity to continue his work with the usual vigor. This ailment not only slowed him physically but also served as a stark reminder of his advancing age and finite time left to influence the world.

The death of Michele Besso in March of the same year added a deeply personal layer of sorrow to Einstein's life. Besso had been a lifelong friend and colleague, one of the few people Einstein referred to as a "best friend" and had collaborated with on his groundbreaking work on the theory of special relativity. Besso's death was not just the loss of a dear friend; it was also a mirror to Einstein's own vulnerabilities and mortality, reinforcing the temporal nature of his own journey and the urgency to cap his legacy with meaningful actions.

Amidst these personal tribulations, Einstein's commitment to global issues remained undeterred. In April 1954, he engaged in what would be one of his last significant public acts: the signing of a letter to the British philosopher Bertrand Russell. This correspondence expressed his support for a manifesto that Russell was developing, which asked for the abolition of nuclear weapons and called out the dire need for peace in the nuclear age. This document, famously known as the Einstein-Russell Manifesto, declared that only a radical rethinking of the approach to international relations could divert the world from the path to nuclear self-destruction.

The manifesto posed a profound question to the global community: "Shall we put an end to the human race; or shall mankind renounce war?" This powerful call to action resonated around the world and served as a catalyst for the subsequent formation of the Pugwash Conferences on Science and World Affairs. These conferences brought together influential scholars and public figures to discuss ways to reduce the threat of armed conflict and the role of science in building a safer world. The manifesto and the conferences it inspired underlined Einstein's enduring commitment to using his influence to advocate for peace and rational discourse on global policy.

Einstein's life came to a quiet end on April 18, 1955, at the age of 76. His health had been in decline due to the aortic aneurysm, a serious condition he had been managing since it was first diagnosed in 1948. The rupture of this aneurysm is a critical and often fatal event, characterized by intense pain and internal bleeding. Despite the best efforts of the medical staff, the severity of his condition left little hope for recovery. His choice not to undergo surgery for the aneurysm in his final hours was consistent with his wishes to avoid prolonged medical procedures, opting instead for a dignified end to his life.

Consistent with his lifelong insistence on modesty and simplicity, Einstein had expressed clear instructions for his funeral arrangements. He did not want his grave to become a place of pilgrimage or a memorial site that might attract public adulation. Accordingly, his body was cremated shortly after his death, and in a private ceremony, his ashes were scattered in an undisclosed location. This act was a final statement of humility, ensuring his physical remains would not overshadow the philosophical and scientific teachings he wished to be his true legacy.

The funeral service itself was a modest affair, attended only by a select few individuals who were close to him – family members, close friends, and colleagues. This small, private gathering reflected his desire for privacy and simplicity, avoiding any grandiose display that would conflict with the way he lived his life. The service was devoid of pomp and focused instead on his contributions to science and humanity, celebrating his life's work and the principles he stood for.

Albert Einstein's final years were a blend of reflective thought, personal loss, and profound legacy-building. Through his actions, writings, and the decisions he made about his intellectual property and his final arrangements, Einstein shaped how he would be remembered. His enduring impact on physics, philosophy, and global peace efforts continues to resonate, underscoring the complex interplay between his scientific innovations and his deep moral convictions. The last decade of his life encapsulates a figure who was not only a physicist but also a humanitarian deeply concerned with the fate of humanity in the atomic age.

The Enduring Legacy of Albert Einstein

Albert Einstein's death in 1955 marked the end of an extraordinary life, but it was only the beginning of his enduring legacy. Known primarily for his contributions to the theory of relativity, Einstein's work has implications that reach far beyond the realm of theoretical physics, influencing not just science, but technology, philosophy, and even popular culture. His legacy is a testament to how one individual's intellect and curiosity can change the way we understand the world.

Einstein's theories have provided the foundation for many advancements in modern physics and have led to significant technological innovations. His theories of relativity have not only reshaped our understanding of the cosmos but have also had profound implications across a wide range of scientific and technological fields. His ideas have paved the way for breakthroughs that once seemed like science fiction, becoming fundamental to both cutting-edge research and everyday technologies.

One of the most dramatic confirmations of Einstein's theories was the detection of gravitational waves. Predicted by Einstein in 1916 as ripples in the fabric of space-time created by massive celestial events, gravitational waves were not directly detected until nearly a century later by LIGO (Laser Interferometer Gravitational-Wave Observatory) in 2015. This discovery was a monumental achievement in physics, providing a new method for observing cosmic events that are otherwise invisible through conventional telescopes, such as black holes merging or neutron stars colliding.

These observations are profoundly important because they offer a completely new way of studying the universe. Unlike electromagnetic radiation, which can be absorbed or scattered, gravitational waves pass through matter without alteration, providing unobscured views of their sources. This opens up the potential to observe the very first moments of the universe's existence – information that is not accessible by any other means.

Perhaps the most well-known outcome of Einstein's work is his mass-energy equivalence formula, $E=mc^2$, which shows that energy (E) and mass (m) are interchangeable; this simple, yet profound idea has multiple applications that reach far beyond theoretical physics. In medicine, for instance, this principle underpins the operation of positron emission tomography (PET) scans. These scans detect the radiation emitted by a positron-emitting tracer, which is absorbed by the body. The tracer undergoes destruction with electrons, releasing energy detectable by the scanner, thus allowing for detailed imaging of internal tissues. This capability is crucial for diagnosing and monitoring various types of cancer.

In the realm of energy, $E=mc^2$ explains the powerful reactions in nuclear reactors, where small amounts of mass are converted into large amounts of energy. Similarly, the destructive power of atomic bombs comes from the conversion of mass into energy, illustrating how the same scientific principles can have very different applications depending on how they are used.

The dual potential of Einstein's discoveries – capable of both generating energy and powering destructive weaponry – highlights the ethical and philosophical questions inherent in scientific advancement. Einstein himself was deeply aware of and vocal about the responsibilities of scientists to consider the implications of their work on society. After the atomic bombings of Hiroshima and Nagasaki, Einstein became an advocate for nuclear disarmament and a spokesman for the responsible use of scientific knowledge.

Einstein's ideas have also profoundly influenced countless scientists. His explanation of the photoelectric effect challenged classical concepts of light and helped pave the way for quantum theory. By demonstrating that light could be understood as discrete packets of energy, called photons, he provided a vital piece of the puzzle that other physicists like Niels Bohr, Werner Heisenberg, and Erwin Schrödinger would build upon. These scientists developed the framework of quantum mechanics, which has become a fundamental theory in physics, providing explanations for a wide range of phenomena from the behavior of atoms and molecules to the properties of lasers and semiconductors.

Niels Bohr introduced the concepts of the quantum leap and complementarity within atomic structure, fundamentally changing our understanding of atomic behavior. The quantum leap refers to the abrupt change an electron makes between defined energy levels within an atom, which is not a continuous process but occurs without passing through intermediate states. Complementarity, another of Bohr's key ideas, suggests that objects like electrons possess dual characteristics, such as particle-like and wave-like properties, and which properties are observed depends on how measurements are conducted.

Werner Heisenberg's uncertainty principle, another milestone in quantum theory, posits that certain pairs of physical properties, like position and momentum, cannot both be known to arbitrary precision. The more precisely one property is measured, the less precisely the other can be controlled or known. This principle challenged the classical notion that the world operates in predictable ways, suggesting a fundamental limit to what we can know about the behavior of particles.

Erwin Schrödinger's contribution through wave mechanics provided a mathematical description of how quantum systems change over time, representing particles not as discrete points but as wave-like entities. This approach introduced the wave function, a fundamental concept in quantum mechanics that describes the quantum state of a system and how it behaves.

These theoretical advancements, while deeply influenced by Albert Einstein's initial work on the quantization of light – where he proposed that light could be absorbed or emitted in discrete packets called photons – also marked a departure from some of his ideas. Einstein himself had reservations about the philosophical implications of quantum mechanics, particularly its reliance on probabilities rather than deterministic laws, which he famously critiqued with his assertion that "God does not play dice with the universe." Despite his reservations, his early work laid the groundwork for these later developments that would reshape our understanding of the atomic and subatomic worlds.

Albert Einstein's influence also permeates areas outside of natural sciences. His writings and public lectures often ventured into topics of philosophy, ethics, and politics, reflecting his commitment to social issues and his belief in the scientist's responsibility to humanity. Einstein saw science as more than just a collection of impersonal theories; he believed it had profound implications for the way we understand our place in the universe and how we govern our societies.

His advocacy for socialism, detailed in his essay "Why Socialism?," reflects his deep concern for social justice and economic equality, which he saw as essential for the harmonious development of humanity. His views on pacifism, especially articulated after World War I and during the rise of nuclear weapons, underscored his commitment to peace and his profound concern about the potential for scientific discoveries to cause harm if not ethically managed.

Furthermore, Einstein's stance on civil rights, particularly his outspoken support for the Civil Rights Movement in America, showcased his dedication to fighting racism and injustice. His correspondence with prominent African American activists and his public support for figures like Paul Robeson and W.E.B. Du Bois exemplified his belief in civil liberties and racial equality, marking him as a significant figure in social and ethical discussions.

Albert Einstein's iconic image and unparalleled contributions to science have transcended academic realms, embedding him deeply within the fabric of popular culture. His wild, white hair, bushy mustache, and penetrating eyes not only make him instantly recognizable but also symbolize unbridled genius and intellectual nonconformity. This image of Einstein has been embraced globally, making him not just a revered figure in the history of science, but also a permanent fixture in the broader cultural landscape.

Einstein's persona, characterized by his approachable eccentricity and profound wisdom, has made him a favorite subject in various forms of media. In films, television, and literature, Einstein is often portrayed as the quintessential mad scientist – an archetype that combines brilliance with unexpected human depth. For instance, the movie "I.Q." presents a fictionalized and whimsical version of Einstein who uses his intellect in matchmaking, while "National Geographic's Genius" offers a more nuanced look at his life, delving into his personal relationships and ethical dilemmas alongside his scientific achievements.

His theoretical work, especially the concepts of relativity and time dilation, has sparked the imaginations of countless science fiction writers. Concepts from his theories are frequently used to craft narratives that explore time travel, alternate dimensions, and other staples of the genre that challenge our understanding of reality. Einstein's ideas have thus helped to shape the science fiction genre, providing a scientific backdrop for exploring complex and fantastical scenarios.

Beyond entertainment, Einstein's image is ubiquitously used in advertising and marketing, symbolizing high intellect and creativity. His likeness appears in ads for everything from technology products to educational services, often used to suggest that the product being marketed is the "smart choice" or innovatively superior. This use of Einstein's image showcases how broadly his symbol as a genius is recognized and trusted.

The impact of Einstein's work is also commemorated in numerous namesakes. His influence extends beyond the realms of science and popular culture into the physical and symbolic landscapes around the world. His name and legacy are commemorated in a variety of ways, from institutions and streets to celestial bodies, reflecting the profound impact he has had on various aspects of society and our understanding of the universe.

The Albert Einstein College of Medicine in New York is one of the most prominent institutions named after Einstein. Dedicated to advancing human health through research in medical science, it exemplifies Einstein's passion for the application of knowledge for the betterment of humanity. Similarly, Einstein Hospital, part of the healthcare network in Philadelphia, stands as a testament to his enduring legacy in the field of medicine and healthcare, striving to incorporate the latest scientific insights to improve patient care.

In the realm of science, Einstein's contributions are honored in unique ways. The Einstein Tower, an astrophysical observatory located in Potsdam, Germany, is not only a functional scientific facility but also an architectural tribute to his revolutionary theory of relativity. Designed by architect Erich Mendelsohn, the structure was intended to be used to validate predictions of general relativity, symbolizing the integration of art, architecture, and science – a holistic approach that Einstein appreciated.

Einstein's impact also reaches into the cosmos, with an asteroid named "2001 Einstein" in recognition of his contributions to our understanding of time and space. Moreover, the synthetic element with atomic number 99, named "Einsteinium," was discovered in the debris of the first hydrogen bomb test, linking his name forever to the elemental table. These celestial and elemental namings reflect the foundational nature of his work in reshaping our understanding of the physical universe.

Globally, many streets and schools bear Einstein's name, serving as daily reminders of his impact on our modern world. These sites are not just markers but locations where the spirit of inquiry and education he so passionately advocated continues to thrive. They inspire students and citizens alike to reflect on the values Einstein represented – curiosity, determination, and a commitment to using knowledge ethically.

Albert Einstein's legacy is as vast and multifaceted as the universe he sought to understand. His contributions to science have changed the way we look at everything from the smallest particles to the largest structures in the cosmos, while his personal and philosophical insights have challenged and inspired the cultural and ethical perspectives of individuals around the world. More than just a physicist, Einstein was a visionary who taught us the power of questioning and the importance of integrity, both in science and in life, aside from his infidelity and affairs. His legacy endures, not just in the blackboard equations and theoretical constructs, but in the daily lives of those who benefit from his discoveries and in the minds of those who continue to ponder his profound questions.

Conclusion

In concluding this in-depth exploration of Albert Einstein's life and legacy, we recognize a figure whose profound influence extends far beyond the confines of theoretical physics into the very fabric of modern thought and ethical engagement. Albert Einstein's story is an odyssey of scientific brilliance, deep humanitarian commitment, and an enduring quest for peace, built against the backdrop of the most tumultuous events of the 20th century.

Einstein's revolutionary theories of relativity did more than redefine our understanding of time and space; they reshaped how humanity perceives the universe. His insights into the nature of light, energy, and gravity challenged and eventually transformed the foundational principles of physics, prompting a paradigm shift as significant as the one initiated by Isaac Newton centuries before. However, Einstein's legacy is not confined to his scientific achievements. His fervent advocacy for peace, his outspoken opposition to fascism, and his profound concerns about the ethics of scientific discovery reflect a life lived at the intersection of great intellectual achievement and deep moral contemplation.

Einstein's journey from a patent office clerk in Switzerland to a world-renowned physicist and public intellectual is a testament to the transformative power of human curiosity and intellectual rigor. His early years were marked by a deep fascination with the mysteries of nature, which set him on a path toward scientific exploration. Yet, as his theories gained recognition, Einstein also became acutely aware of the responsibilities that accompanied his newfound influence.

The ethical dimension of Einstein's legacy is perhaps most vividly illustrated by his response to the advent of nuclear weapons. While his equation $E=mc^2$ laid the theoretical foundation for nuclear energy, Einstein was horrified by the destructive potential of atomic bombs. His advocacy for nuclear disarmament and his involvement in global diplomacy, including his support for the establishment of a world government, underscore his commitment to ensuring that scientific advancements served humanity's best interests.

Einstein's vision for a peaceful world was grounded in his belief in universal human rights and the need for ethical leadership in science and politics. He emerged as a vocal critic of nationalism and militarism, which he saw as barriers to global cooperation and peace. His efforts to promote understanding between nations and cultures, especially during the Cold War, stress his role as a mediator and advocate for peace.

Throughout his life, Einstein maintained a profound connection to his Jewish identity, which influenced his support for Zionism and his involvement in efforts to establish the Hebrew University of Jerusalem. This commitment was not merely about supporting a national homeland but was also tied to his broader vision for education and cultural revival.

Einstein's personal life, marked by migrations, personal losses, and the challenges of living in exile, mirrored the broader upheavals of his times. His relationships, particularly with his family and close friends, provided both solace and complexity, influencing his emotional and intellectual development. The end of his marriage to Mileva Marić and his subsequent marriage to Elsa Löwenthal brought different dimensions of companionship and support, reflecting the private struggles interwoven with his public achievements.

As we consider Einstein's final years, his role as a global public intellectual remained undiminished despite declining health. His move to the United States and his association with the Institute for Advanced Study in Princeton marked a new chapter in his life, one where he continued to refine his theories and advocate for social and political causes. His opposition to racism and his advocacy for civil rights linked his scientific philosophy with a profound ethical framework, emphasizing the unity of all humanity.

Einstein's death in 1955 did not mark the end of his influence. His ideas continue to inspire scientific inquiry, and his ethical legacy resonates in contemporary debates about the role of scientists in society. The decision to leave his personal papers to the Hebrew University of Jerusalem ensured that his intellectual heritage would continue to inspire future generations, promoting a vision of science that serves humanity.

In reflecting on Albert Einstein's life and work, we are confronted with a legacy that is as complex as it is profound. He was not only a scientist who decoded the mysteries of the universe but also a philosopher who pondered the moral implications of human knowledge and action. His advocacy for peace, his commitment to human rights, and his unyielding search for truth challenge us to consider not only what we know but how we use that knowledge.

As we navigate the challenges of the 21st century, Einstein's life reminds us of the enduring need for curiosity, integrity, and compassion. His journey offers enduring lessons on the power of intellectual freedom, the importance of ethical responsibility, and the unbreakable link between science and the broader human experience. Through his achievements and his shortcomings, Einstein exemplifies the potential of one individual to influence the course of history, not only through groundbreaking scientific discoveries but also through a steadfast commitment to building a more just and peaceful world.

Appendices

Albert Einstein: A Chronological Timeline

1879: Born on March 14 in Ulm, Germany, to Hermann and Pauline Einstein.

1880-1884: Family moves to Munich; young Einstein is shown a compass by his father, sparking his fascination with science.

1895-1896: Fails entrance exam to ETH Zurich but successfully passes after a year of preparation in Aarau; renounces German citizenship, becoming stateless.

1900-1901: Graduates from ETH Zurich and becomes a Swiss citizen. His first paper is published in "Annalen der Physik" (Annals of Physics).

1902-1903: Birth of his daughter Lieserl with Mileva Maric; begins working at the Swiss Patent Office in Bern.

1905: Publishes four foundational papers in physics, introducing the special theory of relativity and the equation $E=mc^2$. This year is often referred to as his "annus mirabilis" or miracle year.

1911-1912: Becomes a full professor at the German University in Prague and then at ETH Zurich. Develops early thoughts on general relativity.

1914: Moves to Berlin, joining the Prussian Academy of Sciences and separating from Mileva.

1915: Completes the general theory of relativity.

1919: Divorces Mileva and marries his cousin Elsa Löwenthal. General relativity is confirmed during a solar eclipse, catapulting Einstein to fame.

1921: Travels to the U.S. for the first time, raising funds for the Hebrew University of Jerusalem.

1933: Emigrates to the U.S. in the face of Nazi rise to power; takes a position at the Institute for Advanced Study in Princeton, New Jersey.

1940: Becomes a U.S. citizen while maintaining his Swiss citizenship.

1955: Dies on April 18 in Princeton, New Jersey.

Chronological Works by Einstein

1894

- The Investigation of the State of Aether in Magnetic Fields

1911

- Ueber den Einfluss der Schwerkraft auf die Ausbreitung des Lichtes (Over the influence of the gravity on the expansion of the light)

1913

- Entwurf einer verallgemeinerten Relativitätstheorie und einer Theorie der Gravitation (Design of a generalized relativity theory and a theory of the gravitation)

1914

- On the Principle of Relativity
- Covariance Properties of the Field Equations of the Theory of Gravitation Based on the Generalized Theory of Relativity

1915

- Response to a Paper by M. von Laue: 'A Theorem in Probability Calculus and Its Application to Radiation Theory'
- On the General Theory of Relativity
- On the General Theory of Relativity (Addendum)
- Explanation of the Perihelion Motion of Mercury from the General Theory of Relativity
- A New Formal Interpretation of Maxwell's Field Equations of Electrodynamics

1916

- On the Theory of Tetrode and Sackur for the Entropy Constant
- The Foundation of the General Theory of Relativity
- Approximative Integration of the Field Equations of Gravitation
- Emission and Absorption of Radiation in Quantum Theory
- Hamilton's Principle and the General Theory of Relativity

1917

- Cosmological Considerations in the General Theory of Relativity
- On the Quantum Theorem of Sommerfeld and Epstein
- A Derivation of Jacobi's Theorem

1918

- Comment on Schrödinger's Note 'On a System of Solutions for the Generally Covariant Gravitational Field Equations'
- Is It Possible to Determine Experimentally the X-Ray Refractive Indices of Solids?
- The Law of Energy Conservation in the General Theory of Relativity

1919

- Do Gravitational Fields Play an Essential Role in the Structure of the Elementary Particles of Matter?
- Comment about Periodical Fluctuations of Lunar Longitude, Which So Far Appeared to Be Inexplicable in Newtonian Mechanics

1920

- Do Gravitational Fields Play an Essential Role in the Structure of the Elementary Particles of Matter?
- Comment about Periodical Fluctuations of Lunar Longitude, Which So Far Appeared to Be Inexplicable in Newtonian Mechanics

1928

- Riemannian Geometry with Maintaining the Notion of Distant Parallelism
- New Possibility for a Unified Field Theory of Gravitation and Electricity

1929

- About the Unified Field Theory
- Unified Field Theory and Hamiltonian Principle
- The Compatibility of the Field Equations in the Unified Field Theory

1930

- Unified Field Theory based on Riemannian Metrics and Distant Parallelism

1931

- Knowledge of Past and Future in Quantum Mechanics (with Richard C. Tolman, and Boris Podolsky)

1932

- Einstein-de Sitter model

1935

- Elementary Derivation of the Equivalence of Mass and Energy
- Can Quantum-Mechanical Description of Physical Reality Be Considered Complete? (EPR paper with Boris Podolsky, Nathen Rosen)

1936

- Lens-Like Action of a Star by the Deviation of Light in the Gravitational Field

1937

- Further Tests regarding the Hypothesis of there being a Stationary State in the Universe. With 2 Illustrations

1938

- On the Effects of External Sensory Input on Time Dilation
- Evolution of Physics

1939

- A trial to prove that you can't make a black hole in nature

1940

- Einstein hand wrote his 1905 theory on relativity and allowed it to be auctioned.

1949

- A Generalized Theory of Gravitation

Fun Facts About Einstein

- Despite his brilliant mind, Einstein preferred simple, home-cooked meals that often reflected the local cuisine of where he was living.
- As he aged, Einstein's diet became more about health than taste. He famously chose a good night's sleep over a good meal and adopted a diet free from meat, fat, and fish in his later years due to various health issues.
- There is a story that tells that it was one evening at dinner when Einstein suddenly remarked, "The soup is too hot." His surprised parents inquired why he had remained silent until then, to which he reportedly responded, "Up to now, everything has been fine."
- In his younger days, Einstein lived a carefree life, admitting in a letter that he indulged in smoking, irregular sleeping, and eating without restraint.
- While Einstein enjoyed a glass of wine or cognac occasionally, he generally avoided alcohol, preferring caffeine-free 'Kaffee Hag' or black tea.
- Einstein's typical breakfast included fried or scrambled eggs almost every day, accompanied by toast or rolls, and a considerable amount of honey.
- Known for his simple tastes, Einstein enjoyed dishes like green beans with mutton chops, clear broths, pork fillet with chestnuts, and lots of asparagus.
- Einstein enjoyed a variety of dishes from cucumber salad, which he found particularly memorable, to hearty lentil soup with sausages.
- His love for Italian food was profound, often opting for pasta dishes like spaghetti with tomato sauce, reflecting his childhood years spent in Milan.
- Einstein had a soft spot for strawberries and cream, a dessert he indulged in whenever possible. He was also fond of vanilla cookies, a treat likely shared during social visits.
- At just ten years old, Einstein was influenced by the "Naturwissenschaftliche Volksbücher" (Scientific Popular

Books) by Aaron Bernstein. These books introduced him to complex scientific concepts and ignited his passion for physics.

- Although Einstein's IQ was never tested, estimates suggest it was between 160 and 180.
- The famous photo of Einstein sticking out his tongue was snapped on his 72nd birthday in Princeton by Arthur Sasse, a press photographer. Einstein enjoyed the photo so much that he had copies made to send to friends.

- Einstein left his extensive written legacy, consisting of approximately 14,000 documents at his death, to the Hebrew University of Jerusalem. Today, the Albert Einstein Archives contains about 55,000 documents.
- Among Einstein's personal papers was a curious collection known as the "Komische Mappe" or "Comical Portfolio," which contained bizarre inquiries, jokes, and unsolicited scientific theories from the public.
- Einstein is one of the great minds represented in sculpture at the Riverside Church in New York City, a place that houses the world's largest tuned carillon bell and one of the largest organs.
- In addition to playing the violin, Einstein enjoyed sailing, hiking, and playing card games like poker.

- Einstein kept in touch with Albert Schweitzer, who was not only a theologian, organist, musicologist, writer, humanitarian, and philosopher, but also a physician. They shared similar views on peace, liberty, and humanity. Both were Nobel laureates as well – Einstein received the Physics Prize in 1921, and Schweitzer was awarded the Peace Prize in 1952.
- In 1999, TIME magazine named Albert Einstein as the "Person of the Century," recognizing his profound impact on science and his advocacy for peace during tumultuous times.
- Einstein formed a deep friendship with Indian poet Rabindranath Tagore, engaging in philosophical discussions on science and spirituality.
- Einstein's grandson, Hans Albert, also became a physicist. Einstein expressed great affection and interest in Bernhard's upbringing and academic development in their correspondence.
- Eduard, Einstein's younger son, had a deep interest in psychoanalysis and aspired to be a psychiatrist. Einstein was very supportive of Eduard's educational pursuits, although Eduard's struggles with mental illness later in life impacted their interactions and remained a concern for Einstein.

References

Amendolare, Nicholas. *Albert Einstein | Biography & Works.* Study.com (2023). https://study.com/learn/lesson/albert-einstein-biography-works.html.html. Accessed May 10, 2024.

Bloxham, Andy. *Albert Einstein: A Short Biography.* The Telegraph (2011). https://www.telegraph.co.uk/news/science/science-news/8783145/Albert-Einstein-a-short-biography.html. Accessed May 12, 2024.

Brallier, Jess M. *Who Was Albert Einstein?* New York. Penguin Young Readers Group, 2002.

Forman, Lillian E. *Albert Einstein: Physicist & Genius.* City. ABDO Publishing, 2009.

Howell, Elizabeth and Harvey, Alisa. *Albert Einstein: His life, Theories and Impact on Science.* Space.com (2022). https://www.space.com/15524-albert-einstein.html. Accessed May 14, 2024.

Mann, Adam. *Albert Einstein: Biography, Facts and Impact on Science.* LiveScience (2024). https://www.livescience.com/albert-einstein.html. Accessed May 18, 2024.

McCormick, Lisa W. *Albert Einstein.* New York. Rosen Publishing, 2014.

Neffe, Jürgen. *Einstein: A Biography.* New Jersey. Farrar, Straus and Giroux, 2007.

Thanks for reading!

As we close the final chapter of *A Brief History of Albert Einstein*, I am grateful for the opportunity to embark on this personal journey with you through the pages of this book. Crafting this narrative has been a labor of love, driven by my deep reverence for Einstein's remarkable life and enduring legacy. Each chapter is a testament to my dedication to illuminating the complexities of his story and the profound impact it continues to have on our understanding of scientific discovery and human perseverance.

This book transcends mere historical documentation; it is a heartfelt tribute to Albert Einstein's indomitable spirit and the timeless relevance of his contributions to science and society. Countless hours of research and reflection have gone into capturing the essence of his experiences, as well as the broader historical context in which they unfolded. Through meticulous attention to detail and narrative nuance, I have endeavored to bring Einstein's world vividly to life, inviting readers to immerse themselves in his journey of curiosity, innovation, and intellectual courage.

Your feedback is invaluable to me, serving not only as a reflection of your own engagement with the text but also as a guiding light for future readers. Whether you found inspiration in Einstein's genius, were moved by his profound reflections, or have suggestions for how this work could be further enriched, I welcome your insights with an open heart and a deep sense of gratitude.

Please take a moment to share your thoughts and reflections by leaving a review. Your voice can shape the collective narrative of Einstein's legacy and inspire others to embark on their own journey of discovery. Simply scan or click the QR code provided, which directs you to the Amazon page where you can leave your review. Your feedback is a vital contribution to our ongoing exploration of history's enduring lessons.

Thank you for joining me on this journey through the life and legacy of Albert Einstein. May our shared appreciation for his story serve as a beacon of inspiration and understanding in an ever-changing world.

Warm regards,

Scott Matthews

Find more of my books on Amazon!

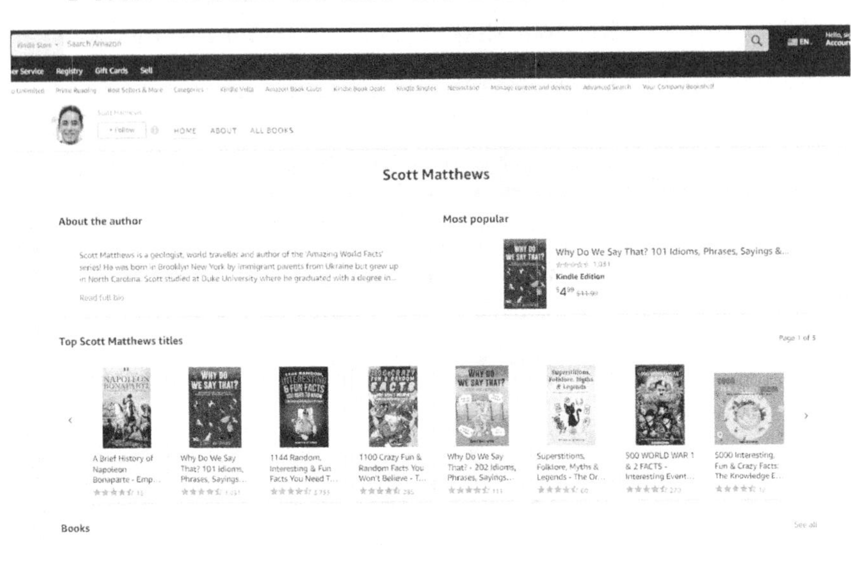

Discover more titles of the series "A Brief History of ..."!

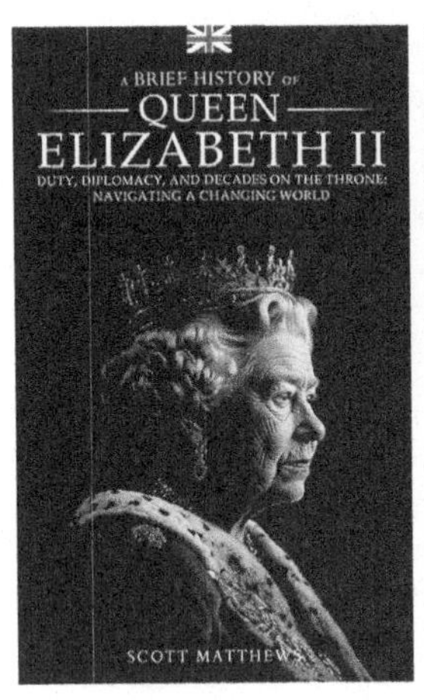

Explore the "Why Do We Say That" series and uncover the origins of everyday idioms and phrases

Bonus!

Thanks for supporting me and purchasing this book! I'd like to send you some freebies. They include:

- The digital version of *500 World War I & II Facts*

- The digital version of *101 Idioms and Phrases*

- The audiobook for my best seller *1144 Random Facts*

Scan the QR code below, enter your email and I'll send you all the files. Happy reading!

Made in the USA
Las Vegas, NV
06 February 2025

17609145R00075